Lecture Notes in Mathematics

For information about Vols. 1–551, please contact your bookseller or Springer-Verlag.

Vol. 552: C. G. Gibson, K. Wirthmüller, A. A. du Plessis and E. J. N. Looijenga. Topological Stability of Smooth Mappings. V, 155 pages. 1976.

Vol. 553: M. Petrich, Categories of Algebraic Systems. Vector and Projective Spaces, Semigroups, Rings and Lattices. VIII, 217 pages. 1976.

Vol. 554: J. D. H. Smith, Mal'cev Varieties. VIII, 158 pages. 1976.

Vol. 555: M. Ishida, The Genus Fields of Algebraic Number Fields. VII, 116 pages. 1976.

Vol. 556: Approximation Theory. Bonn 1976. Proceedings. Edited by R. Schaback and K. Scherer. VII, 466 pages. 1976.

Vol. 557: W. Iberkleid and T. Petrie, Smooth S^1 Manifolds. III, 163 pages. 1976.

Vol. 558: B. Weisfeiler, On Construction and Identification of Graphs. XIV, 237 pages. 1976.

Vol. 559: J.-P. Caubet, Le Mouvement Brownien Relativiste. IX, 212 pages. 1976.

Vol. 560: Combinatorial Mathematics, IV, Proceedings 1975. Edited by L. R. A. Casse and W. D. Wallis. VII, 249 pages. 1976.

Vol. 561: Function Theoretic Methods for Partial Differential Equations. Darmstadt 1976. Proceedings. Edited by V. E. Meister, N. Weck and W. L. Wendland. XVIII, 520 pages. 1976.

Vol. 562: R. W. Goodman, Nilpotent Lie Groups: Structure and Applications to Analysis. X, 210 pages. 1976.

Vol. 563: Séminaire de Théorie du Potentiel. Paris, No. 2. Proceedings 1975–1976. Edited by F. Hirsch and G. Mokobodzki. VI, 292 pages. 1976.

Vol. 564: Ordinary and Partial Differential Equations, Dundee 1976. Proceedings. Edited by W. N. Everitt and B. D. Sleeman. XVIII, 551 pages. 1976.

Vol. 565: Turbulence and Navier Stokes Equations. Proceedings 1975. Edited by R. Temam. IX, 194 pages. 1976.

Vol. 566: Empirical Distributions and Processes. Oberwolfach 1976. Proceedings. Edited by P. Gaenssler and P. Révész. VII, 146 pages. 1976.

Vol. 567: Séminaire Bourbaki vol. 1975/76. Exposés 471–488. IV, 303 pages. 1977.

Vol. 568: R. E. Gaines and J. L. Mawhin, Coincidence Degree, and Nonlinear Differential Equations. V, 262 pages. 1977.

Vol. 569: Cohomologie Etale SGA 4½. Séminaire de Géométrie Algébrique du Bois-Marie. Edité par P. Deligne. V, 312 pages. 1977.

Vol. 570: Differential Geometrical Methods in Mathematical Physics, Bonn 1975. Proceedings. Edited by K. Bleuler and A. Reetz. VIII, 576 pages. 1977.

Vol. 571: Constructive Theory of Functions of Several Variables, Oberwolfach 1976. Proceedings. Edited by W. Schempp and K. Zeller. VI, 290 pages. 1977

Vol. 572: Sparse Matrix Techniques, Copenhagen 1976. Edited by V. A. Barker. V, 184 pages. 1977.

Vol. 573: Group Theory, Canberra 1975. Proceedings. Edited by R. A. Bryce, J. Cossey and M. F. Newman. VII, 146 pages. 1977.

Vol. 574: J. Moldestad, Computations in Higher Types. IV, 203 pages. 1977.

Vol. 575: K-Theory and Operator Algebras, Athens, Georgia 1975. Edited by B. B. Morrel and I. M. Singer. VI, 191 pages. 1977.

Vol. 576: V. S. Varadarajan, Harmonic Analysis on Real Reductive Groups. VI, 521 pages. 1977.

Vol. 577: J. P. May, E_∞ Ring Spaces and E_∞ Ring Spectra. IV, 268 pages. 1977.

Vol. 578: Séminaire Pierre Lelong (Analyse) Année 1975/76. Edité par P. Lelong. VI, 327 pages. 1977.

Vol. 579: Combinatoire et Représentation du Groupe Symétrique, Strasbourg 1976. Proceedings 1976. Edité par D. Foata. IV, 339 pages. 1977.

Vol. 580: C. Castaing and M. Valadier, Convex Analysis and Measurable Multifunctions. VIII, 278 pages. 1977.

Vol. 581: Séminaire de Probabilités XI, Université de Strasbourg. Proceedings 1975/1976. Edité par C. Dellacherie, P. A. Meyer et M. Weil. VI, 574 pages. 1977.

Vol. 582: J. M. G. Fell, Induced Representations and Banach *-Algebraic Bundles. IV, 349 pages. 1977.

Vol. 583: W. Hirsch, C. C. Pugh and M. Shub, Invariant Manifolds. IV, 149 pages. 1977.

Vol. 584: C. Brezinski, Accélération de la Convergence en Analyse Numérique. IV, 313 pages. 1977.

Vol. 585: T. A. Springer, Invariant Theory. VI, 112 pages. 1977.

Vol. 586: Séminaire d'Algèbre Paul Dubreil, Paris 1975–1976 (29ème Année). Edited by M. P. Malliavin. VI, 188 pages. 1977.

Vol. 587: Non-Commutative Harmonic Analysis. Proceedings 1976. Edited by J. Carmona and M. Vergne. IV, 240 pages. 1977.

Vol. 588: P. Molino, Théorie des G-Structures: Le Problème d'Equivalence. VI, 163 pages. 1977.

Vol. 589: Cohomologie l-adique et Fonctions L. Séminaire de Géométrie Algébrique du Bois-Marie 1965–66, SGA 5. Edité par L. Illusie. XII, 484 pages. 1977.

Vol. 590: H. Matsumoto, Analyse Harmonique dans les Systèmes de Tits Bornologiques de Type Affine. IV, 219 pages. 1977.

Vol. 591: G. A. Anderson, Surgery with Coefficients. VIII, 157 pages. 1977.

Vol. 592: D. Voigt, Induzierte Darstellungen in der Theorie der endlichen, algebraischen Gruppen. V, 413 Seiten. 1977.

Vol. 593: K. Barbey and H. König, Abstract Analytic Function Theory and Hardy Algebras. VIII, 260 pages. 1977.

Vol. 594: Singular Perturbations and Boundary Layer Theory, Lyon 1976. Edited by C. M. Brauner, B. Gay, and J. Mathieu. VIII, 539 pages. 1977.

Vol. 595: W. Hazod, Stetige Faltungshalbgruppen von Wahrscheinlichkeitsmaßen und erzeugende Distributionen. XIII, 157 Seiten. 1977.

Vol. 596: K. Deimling, Ordinary Differential Equations in Banach Spaces. VI, 137 pages. 1977.

Vol. 597: Geometry and Topology, Rio de Janeiro, July 1976. Proceedings. Edited by J. Palis and M. do Carmo. VI, 866 pages. 1977.

Vol. 598: J. Hoffmann-Jørgensen, T. M. Liggett et J. Neveu, Ecole d'Eté de Probabilités de Saint-Flour VI – 1976. Edité par P.-L. Hennequin. XII, 447 pages. 1977.

Vol. 599: Complex Analysis, Kentucky 1976. Proceedings. Edited by J. D. Buckholtz and T. J. Suffridge. X, 159 pages. 1977.

Vol. 600: W. Stoll, Value Distribution on Parabolic Spaces. VIII, 216 pages. 1977.

Vol. 601: Modular Functions of one Variable V, Bonn 1976. Proceedings. Edited by J.-P. Serre and D. B. Zagier. VI, 294 pages. 1977.

Vol. 602: J. P. Brezin, Harmonic Analysis on Compact Solvmanifolds. VIII, 179 pages. 1977.

Vol. 603: B. Moishezon, Complex Surfaces and Connected Sums of Complex Projective Planes. IV, 234 pages. 1977.

Vol. 604: Banach Spaces of Analytic Functions, Kent, Ohio 1976. Proceedings. Edited by J. Baker, C. Cleaver and Joseph Diestel. VI, 141 pages. 1977.

Vol. 605: Sario et al., Classification Theory of Riemannian Manifolds. XX, 498 pages. 1977.

Vol. 606: Mathematical Aspects of Finite Element Methods. Proceedings 1975. Edited by I. Galligani and E. Magenes. VI, 362 pages. 1977.

Vol. 607: M. Métivier, Reelle und Vektorwertige Quasimartingale und die Theorie der Stochastischen Integration. X, 310 Seiten. 1977.

Vol. 608: Bigard et al., Groupes et Anneaux Réticulés. XIV, 334 pages. 1977.

continuation on page 249

Lecture Notes in Mathematics

Edited by A. Dold and B. Eckmann

762

D. H. Sattinger

Group Theoretic Methods in Bifurcation Theory

Springer-veriag
Berlin Heidelberg New York 1979

Author

D. H. Sattinger
School of Mathematics
University of Minnesota
Minneapolis, Minnesota 55455
USA

AMS Subject Classifications (1970): 35 J XX, 35 K XX, 47 H15, 76-XX

ISBN 978-3-540-09715-0 Springer-Verlag Berlin Heidelberg New York
ISBN 978-0-387-09715-2 Springer-Verlag New York Heidelberg Berlin

Library of Congress Cataloging in Publication Data
Sattinger, David H
Group theoretic methods in bifurcation theory.
(Lecture notes in mathematics; 762)
Bibliography: p.
Includes index.
1. Differential equations, Partial--Numerical solutions. 2. Bifurcation theory.
3. Representations of groups. I. Title. II. Series: Lecture notes in mathematics (Berlin); 762.
QA3.L28 no. 762 [QA377] 510'.8s [515'.353] 79-23605
ISBN 978-0-387-09715-2

2141/3140-543210

PREFACE

This set of lectures was given in the winter and spring of 1978 while I was on sabbatical at the University of Chicago. I have tried to present the fundamental ideas involved in the combination of group representation theory and bifurcation theory. In addition, there is a chapter by Peter Olver on the derivation of the symmetry group of a differential equation by algebraic methods.

I would like to thank the Mathematics Department of the University of Chicago for their kind hospitality, Peter Olver for his useful remarks during the lectures and his contribution to these notes, and F. Flowers for an excellent job of typing.

My sabbatical was supported by the University of Minnesota, the National Science Foundation (MCS 73-08535) and the U.S. Army Research Office (DA AG 29-77-G-0122), whose support is much appreciated.

David H. Sattinger
June 1978

TABLE OF CONTENTS

PHYSICAL EXAMPLES OF BIFURCATION

If a layer of fluid is heated from below, convective instabilities set-in when the temperature drop exceeds a certain critical value, and the convective motions which evolve often display a striking cellular structure, as pictured below.

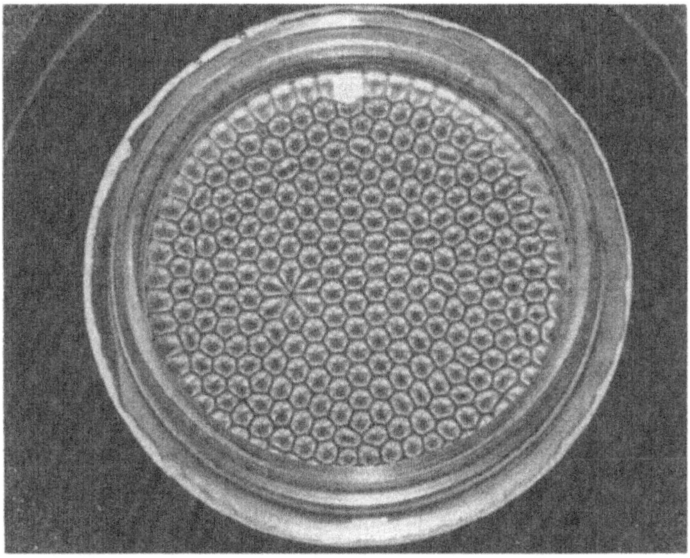

Figure 1.1

Convection cells in the Benard problem [12] +

+ Reprinted from: E. Koschmieder: Benard Convection. Advances in Chemical Physics 26, 177 (1974) by permission of John Wiley and Sons, Inc.

(See Chandrasekhar [4] or Koschmieder [11], [12]) The onset of such convective motions provides a striking physical example of the phenomenon of <u>bifurcation</u> -- that is, the appearance of multiple solutions to a nonlinear system of equations.

To state the situation mathematically, let us suppose that the equilibrium states of a physical (electrical, mechanical, chemical, biological, what have you) system are determined as solutions of a functional equation

$$G(\lambda, u) = 0 \tag{1.1}$$

where λ is a parameter, u is an element of a vector space $\mathcal{E}$ (for example a Banach space) and G is a mapping from $\mathcal{E}$ to another vector space $\mathcal{F}$. We shall give specific examples of such equations later on. Let us suppose that $u = 0$ is always a possible solution of (1.1), that is, $G(\lambda, 0) \equiv 0$. A <u>bifurcation point</u> $(\lambda_c, 0)$ is a critical value of λ at which several solution branches of (1.1) confluesce. For example, there may be a one-parameter family of nontrivial solutions

$$\lambda = \lambda(\epsilon), \quad u = u(\epsilon), \quad G(\lambda(\epsilon), u(\epsilon)) \equiv 0,$$

$$\lambda(0) = \lambda_c \qquad u(0) = 0.$$

The situation is indicated schematically in Figure 1.2 .

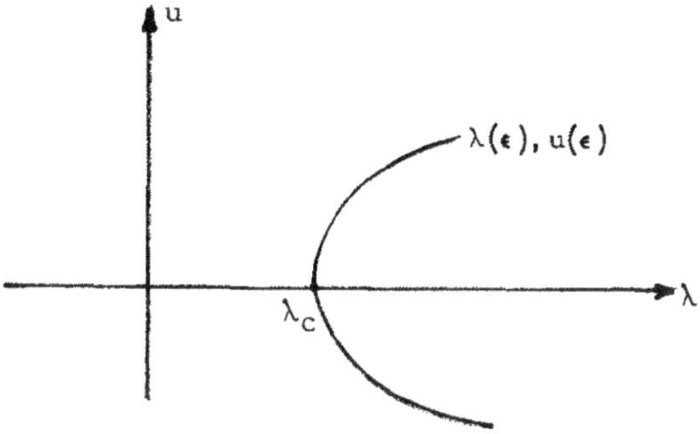

Figure 1.2

Bifurcation point $(\lambda_c, 0)$; one non-trivial branch. The vertical axis represents the vector space $\mathcal{E}$.

Closely tied to the phenomenon of bifurcation is the property of stability. Roughly speaking, a solution is stable if small perturbations from it remain small (We are talking now about the dynamical situation). A solution is asymptotically stable if small disturbances decay to zero in time. When external parameters are varied (such as the temperature drop across the layer of fluid in the Benard problem), one solution may become unstable as the parameter crosses the critical value, and it is at such a transition point that new solutions bifurcate from the known solutions.

The new solutions which appear at the transition to instability need not be stationary solutions. They may be time-dependent solutions — for example time-periodic solutions. In such a case, the loss of stability manifests itself as the appearance of oscillatory motions; generally, the amplitude of these motions increases with the increasing parameter value.

The appearance of time-periodic motions is especially important in the theory of simple mechanical systems (two degrees of freedom) such as clocks and vacuum tube circuits , E. Hopf in 1942 [7] gave a general theorem for the bifurcation of periodic motions in systems with n degrees of freedom. He also discussed the possible extension of his results to systems with an infinite number of degrees of freedom (i.e., systems governed by partial differential equations) and hydrodynamics in particular. Hopf's results have now in fact been generalized to the Navier-Stokes equations (the partial differential equations governing the dynamics of a viscous incompressible fluid) and more generally to parabolic systems of partial differential equations. The generalization to partial differential equations has been discussed in Marsden and McCracken's book [14] and in my lecture notes [18]. I will give a partial derivation of the results in Chapter III of these notes.

The Bénard problem (convection in a plane layer) mentioned above is a model problem which has been much studied in the laboratory because it has given such insight into the processes of bifurcation in fluid dynamics. It is a sensitive problem and many fundamental questions remain unresolved to this day. Of more importance to the physical sciences, however, is the onset of convection in a spherical geometry. These questions have an important bearing on problems in geophysics and astrophysics where the onset of convection in a star or in the earth's core are of interest. We shall discuss these and other bifurcation problems of physical interest below; but the list is only indicative of the variety of bifurcation phenomena which occur in nature.

1. The equations employed to model convection are the so-called Boussinesq equations [4]

$$\Delta u_k + \delta_{k3}\theta - \frac{\partial p}{\partial x_k} = \frac{1}{\mathcal{P}_r} u_j \frac{\partial u_k}{\partial x_j} + \frac{\partial u_k}{\partial t} \tag{1.2a}$$

$$\Delta\theta + \mathcal{R} u_3 = u_j \frac{\partial \theta}{\partial x_j} + \frac{\partial \theta}{\partial t} \tag{1.2b}$$

$$\frac{\partial u_j}{\partial x_j} = 0 \tag{1.2c}$$

where $\mathcal{P}_r = \nu/\kappa$ is the Prandtl number, and $\mathcal{R} = \frac{\alpha g h^3}{\nu\kappa}(T_0 - T_1)$ is the Rayleigh number. Here g is the acceleration due to gravity, h is a characteristic length (for example, the depth of the layer of fluid), α is the coefficient of thermal expansion of the fluid; T_0 and T_1 are the temperatures on the lower and upper boundaries; ν and κ are the coefficients of viscosity and thermal conductivity; θ is the perturbed temperature profile; and u_1, u_2, u_3, p are the components of velocity and pressure. δ_{k3} is the Kronecker delta function.

These same equations are also valid in a spherical geometry. In geophysical applications where the body is rotating, an additional term due to Coriolis forces must be added to the right-side of the first equation.

In vector form this term is $2\Omega \times \vec{u}$ where Ω is the rotation vector ($\epsilon_{k\ell m}\Omega_\ell u_m$ in component form).

One of the outstanding problems in geophysics is the Dynamo problem: How is the earth's magnetic field maintained? The consensus now is that the magnetic field is sustained by electric currents flowing in the electrically conducting earth's core. It is known that purely axi-symmetric fluid motions cannot sustain a magnetic field; but if convective motions accounted for asymmetric fluid motions, these might sustain such an electromagnetic field. The equations of dynamo theory are the convection equations plus an equation for the magnetic field:

$$\frac{\partial \vec{u}}{\partial t} + (\vec{u}\cdot\nabla)\vec{u} + 2\vec{\Omega} \times \vec{u} = -\nabla p - \beta \vec{g}\,\theta + \nu \Delta \vec{u} + \frac{1}{\mu}(\nabla \times \vec{B}) \times \vec{B} \tag{1.3a}$$

$$\operatorname{div} \vec{u} = 0 \tag{1.3b}$$

$$\frac{\partial \theta}{\partial t} + (\vec{u}\cdot\nabla)\theta = -\vec{u}\cdot\nabla T_0 + \kappa \Delta \theta \tag{1.3c}$$

$$\frac{\partial \vec{B}}{\partial t} = \operatorname{curl}(\vec{u} \times \vec{B} + \eta \Delta \vec{B} \tag{1.3d}$$

where T_0 is the base temperature profile in the absence of convection. The term $2\vec{\Omega} \times \vec{u}$ is the Coriolis term due to rotation.

When no magnetic field is present the quadratic term $(\nabla \times \vec{B}) \times \vec{B}$ in (1.3a) vanishes and we have a pure convection problem. When the convective velocity $\vec{u}$ rises to a magnitude and configuration which can sustain a growing magnetic field in (1.3d) then a bifurcation may take place and a non-trivial magnetic field be sustained. The equation (1.3d) is linear so for given $\vec{u}$ one should expect an exponentially growing magnetic field $\vec{B}$; but the nonlinear coupling between $\vec{B}$ and $\vec{u}$ via (1.3a) may be expected to prevent unlimited growth (if the theory is correct), so that a stable equilibrium is attained. Since (1.3d) is homogeneous in $\vec{B}$, some external magnetic field (for example, the sun's) is required to "seed" the dynamo process.

Consideration of (1.3d) without regard to equations (1.3a) through (1.3c) is known as the "kinematic dynamo problem": What velocity fields $\vec{u}$ will sustain a growing magnetic field $\vec{B}$? For further discussion, see Busse [3] and Roberts [17].

2. Bifurcation and Stability in Astrophysics

Stars, as fluid masses, are subject to the classical hydrodynamic instabilities; but, moreover, thermomechanic effects intrinsic to stellar structures can also lead to hydrodynamical effects. In general there is a wide variety of instability mechanisms in stellar structures, and these can lead to bifurcation phenomena in stars. During the course of evolution of a star, one family of equilibrium configurations may become unstable, leading to the bifurcation of secondary family of equilibria. Classically, highly idealized models for stellar structure have been considered. These models suppose the star to be of uniform density and to have an ellipsoidal bounding surface. (The restriction to uniform density is unrealistic for stars.) Assuming the surface is ellipsoidal with semi-axes a_1, a_2, a_3, one must determine these quantities. Since the density ρ is constant, the gravitational potential is given by

$$V(x) = \pi G\rho\{I - \sum_{k=1}^{3} A_k x_k^2\}$$

$$\sum_{k=1}^{3} \frac{x_k^2}{a_k^2} \leq 1 .$$

The coefficients I and A_k depend in a known way on the quantities a_k. Maclaurin showed that an oblate spheroid $(a_2 = a_1 \geq a_3)$ could be a figure of relative equilibrium provided the angular velocity ω around the x_3-axis satisfies

$$\omega^2 = 2\pi G\rho\{A_1 - \frac{a_3^2}{a_1^2} A_3\} \qquad 0 \leq \frac{a_3}{a_1} \leq 1 .$$

In 1834, Jacobi discovered that ellipsoids with three unequal axes are possible provided

$$a_3^2 A_3 = \frac{a_1^2 a_2^2}{a_1^2 - a_2^2} (A_2 - A_1)$$

and

$$\omega^2 = 2\pi G\rho \frac{a_1^2 A_1 - a_2^2 A_2}{a_1^2 - a_2^2} .$$

In the limit as $a_2 \to a_1$, the Jacobi family approaches the Maclaurin family: the Jacobi ellipsoids bifurcate from the Maclaurin spheroids at this point. For a discussion of these and further bifurcation problems in astrophysics, see Lebovitz [13], Michel [15].

3. Bifurcation in Elastic Structures.

There is an extensive literature on bifurcation in elastic structures, that is, buckling of elastic rods and shells. One simple model for the equilibria of an elastic rod is given by the equation

$$[B(s)\theta']' + p\sin\theta = 0 \tag{1.4}$$

where s is the arc length parameter and $\theta(s)$ is the angle between the tangent to the curve at s and the direction of the applied load, and p is the magnitude of the prescribed load; p plays the role of the bifurcation parameter. The function $M(s) = B(s)\theta'(s)$ gives the bending moment M as a function of the curvature θ'. In (1.4) it is linear, but more generally this may be replaced by the nonlinear constitutive relation $M(s) = M(\theta', s)$, leading to the quasilinear equation

$$(M(\theta', s))' + p\sin\theta = 0\ . \tag{1.5}$$

These equations are geometrically exact in that no geometrical approximations are made in the formulation of the equation -- hence the appearance of the sin function. The development of nonlinear plate and shell theory has not been as satisfactory as that of rod theories. One nonlinear plate theory which has received considerable attention in the recent literature is the van Kármán theory developed in 1910. The van Kármán equations are

$$\Delta\Delta w = \lambda[f, w] + [\varphi, w]$$
$$\Delta\Delta\varphi = -[w, w] \qquad (1.6)$$

where Δ is the Laplacian and $[u, v] = u_{xx}v_{yy} + u_{yy}v_{xx} - 2u_{xy}v_{xy}$. These equations are derived on the basis of a number of geometric approximations roughly equivalent to the replacement of $\sin\theta$ by $\theta - \frac{\theta^3}{3!}$. Consequently, the theory is not geometrically exact; and furthermore (1.6) are based on a certain linear stress strain law. For a critique and further discussion of buckling see Antman [1].

4. Biological Applications.

Neural nets are aggregates of nerve cells whose interactions take the form of ordered patterns of electrochemical activity. These organized patterns are presumed to supply the underlying mechanisms for mental processes such as recognition, perception, and memory. At the present time neurons seem either to be excitatory or inhibitory in the effects on other neurons; and they may be interconnected with other cells on a short range basis via axo-somatic, axo-dendritic, or dendro-dentritic synapses (connections) or on a long range basis via insulated myelinated axons. It should be expected that excitatory/inhibitory networks would exhibit such phenomena as bifurcation, threshold effects, and hysteresis. Bifurcation phenomena in simple models of neural nets of excitatory/inhibitory networks have been discussed recently by Cowan and Ermentrout [5]. They consider a simple model neural network governed by the system of equations

$$\mu\dot{Y} = -Y + S(KY + P) \tag{1.7}$$

where Y is a two-component vector, S is a nonlinear vector-valued function, K is a linear convolution operator, and P is the external stimulus (injected current). Equations (1.7) may be studied in one, two, or three dimensions. The convolution operator K arises because the neural activity is spatially averaged (Thus the network of discrete neurons is modeled as a continuum in the limit of high density of neurons). A parameter λ is introduced into the convolution operator K which modifies the output strength of the excitatory cells. With $P = 0$, bifurcation of time independent, spatially periodic solutions of (1.7) can take place as λ crosses a critical value. In physiological situations, the parameter λ may, for example, represent the level of a certain hormone or ionic concentration which modifies the action of the given type of neuron. The presence of periodic wave trains is presumed to represent some kind of short-term memory imprint.

Another area of biology in which bifurcation plays an important role is morphogenesis, as well as in simpler chemically interacting systems. In these problems one tries to explain pattern formation and chemical waves on the basis of bifurcation theory. The kinetics of chemical processes are described by systems of ordinary differential equations

$$\frac{dc_i}{dt} = f_i(c_1, \ldots, c_n, T)$$

$$\frac{dT}{dt} = f_{n+1}(c_1, \ldots, c_n, T)$$

where $c_1, \ldots, c_n$ are concentrations of the various agents and T is the temperature. In some instances, diffusion is important in these processes and then the equations to be considered are parabolic systems. We may write the system generally as

$$\vec{u}_t = K \Delta \vec{u} + \vec{F}(\vec{u}) \tag{1.8}$$

where K is a diagonal matrix whose entries are the diffusion coefficients and $\vec{u} = (c_1, \ldots, c_n, T)$. Due to the competition between reaction and diffusion effects one may observe processes of bifurcation, pattern formation, and wave trains. These areas of investigation have attracted widespread interest in recent years. See Howard [8], Howard and Kopell [9], [10], Fife [6], Othmer [16], and Auchmuty and Nicolis [2]. The idea that morphogenesis in biological structures might be explained on the basis of reaction and diffusion effects goes back to Turing [19].

The equations of chemical kinetics $U_t = F(U)$ normally would have a unique absolutely stable equilibrium representing the ultimate state after all the reactions have run down. The interesting chemical effects occur on a shorter time scale, and to model these reactions it is necessary to make some approximations which produce a simpler set of equations.

For example, some of the concentrations may be regarded as slowly varying and these set equal to a constant. The same approximations are then valid for the reaction-diffusion equations provided these same concentrations are spatially constant.

One may search for special solutions of (1.8), for example, a plane wave solution of the form $U = U(\sigma t - k \cdot x)$ where U is a 2π periodic function and σ and k are a constant scalar and wave vector. The substitution of such a solution in (1.8) leads to the system of ordinary differential equations

$$\sigma U' = F(U) + \alpha^2 K U'' \tag{1.9}$$

where $\alpha = |\hat{k}|$. In general, (1.9) may not have 2π periodic solutions but with certain hypotheses on F, Howard and Kopell have shown that it does [9]. By employing singular perturbation methods (small α) they were able to show the existence of a one parameter family of plane wave solutions for small α under the hypothesis that the kinetic equations $U_t = F(U)$ have a stable limit cycle.

The most thoroughly studied oscillatory chemical reaction reaction is the Belousov reaction. This reaction exhibits self-excited oscillations in which concentrations vary over several orders of magnitude. The period of the oscillation is about 30 seconds and the reaction persists for an hour or two. Thus a two-time perturbation procedure is required. The system exhibits chemical waves.

5. Symmetry Breaking in Elementary Particle Theory

The notion of symmetry breaking plays a fundamental role in elementary particle theory. Such symmetry breaking, viewed from the point of view of bifurcation theory, has been discussed extensively by L. Michel (see, for example, [15]). In Michel's notation the directions of symmetry breaking are obtained as solutions of a quadratic equation over a finite-dimensional vector space: $x \vee x = \lambda x$. The binary operation $\vee$ may be symmetric or anti-symmetric; but in general it is non-associative. The vector solutions x are idempotents of this algebra if $\lambda \neq 0$ and nilpotents if $\lambda = 0$. In the symmetric case such equations are the mathematical equivalent of the reduced bifurcation equations (see Chapter IV). The symmetric bilinear form $Q(x, y)$ arising from the quadratic terms of the bifurcation equations define a suitable product: $x \vee y = Q(x, y)$.

If the original bifurcation problem is covariant with respect to some symmetry group then this covariance is inherited by the bifurcation equations. In elementary particle theories the product is covariant as well: $Tg(x \vee x) = (Tgx) \vee (Tgx)$ where Tg is a linear representation of some symmetry group G. The vector solutions x, however, are not invariant under the entire group G but only under some subgroup; hence the term "symmetry breaking". Some of the symmetry groups which arise in elementary particle theory are $SU(3)$ and $SU(3) \times SU(3)$. In

Chapter 6 we discuss the reduced bifurcation equations $Q(x,x) = \lambda x$ when the symmetry group is SU(2). Such problems arise in classical mechanics (for example convection or buckling problems) when the system is rotationally invariant. But the same equations arise in Gell-Mann's theory of quarks with SU(2) replaced by SU(3), and the representation is the eight-dimensional adjoint representation of SU(3).

Symmetry breaking for the action of SO(n) has been treated in the following way by Michel. Given a symmetric $n \times n$ matrix Q, define $Q = Q - \frac{1}{n}(\mathrm{tr}\ Q)I$; then $\mathrm{tr}\ Q = 0$ and $Q^t = Q$. The inner product defined by $(Q_1, Q_2) = \frac{1}{2}\mathrm{tr}\ Q_1 Q_2$ is invariant under the adjoint action of $SO(n): Q \rightarrow R\,QR^{-1}$. The non-associative product $\vee$ defined by $Q_1 \vee Q_2 = \sqrt{n}/2(Q_1Q_2 + Q_2Q_1) - 2/\sqrt{n}\,(Q_1, Q_2)I$ maps the space of such Q to itself and is covariant with respect to this adjoint action, but it is non-associative. The second and third order functionals $\gamma(Q) = (Q, Q)$ and $\theta(Q) = (Q \vee Q, Q)$ are invariants of the adjoint action. The directions of symmetry breaking are given by solutions to the equation $Q \vee Q = \lambda Q$. These are the Euler-Lagrange equations for the variational problem $\min \theta(Q)$ subject to $\gamma(Q) = 1$. Problems of this type will be discussed in Chapter VI.

References

1. S. Antman, "Bifurcation Problems for Nonlinearly Elastic Structures," in Applications of Bifurcation Theory, Academic Press, New York, 1977.

2. J. F. G. Auchmuty and G. Nicolis, "Dissipative Structures, Catastrophes, and Pattern Formation: A bifurcation analysis," Proc. Nat. Acad. Sci. USA, 71, 2748.

3. F. Busse, "Mathematical Problems of Dynamo Theory," in Applications of Bifurcation Theory, op. cit.

4. S. Chandrasekhar, Hydordynamic and Hydromagnetic Stability, Oxford, Clarendon Press (1961).

5. J. D. Cowan and G. B. Ermentrout, "Secondary Bifurcation in Neuronal Nets," SIAM Jour. Appl. Math.; "A mathematical theory of visual hallucination patterns," Biological Cybernetics.

6. P. Fife, "Pattern formation in reacting and diffusing systems," J. Chem. Phys. 64, 554.

7. E. Hopf, "Abzweigung einer periodischen lösung eines differential Systems," Berichten der Math-Phys. Klasse der Sächischen Akademie der Wissenschaften zu Leipzig, XCIV (1942), pp 1-22.

8. L. N. Howard, "Chemical Wave Trains and Related Structures," in Applications op. cit.

9. L. N. Howard and N. Kopell, "Plane Wave Solutions to Reaction-Diffusion Equations," Studies in Applied Math. 52 (1973) pp. 291-328.

10. L.N. Howard and N. Kipell, "Bifurcations and Trajectories Joining Critical Points," Advances in Math. 18 (1975) pp. 306-358.

11. E.L. Koschmieder, "On Convection of a Uniformly Heated Plane," Beiträge zur Physik der Atmosphäre 39 (1966).

12. __________, "Benard Convection," Advances in Chemical Physics, 26 177 (1974).

13. N. Lebovitz, "Bifurcation and Stability Problems in Astrophysics," in Applications op. cit.

14. J. Marsden and M. McCracken, "The Hopf Bifurcation and its Applications," Applied Mathematical Sciences 19, Springer Verlag, New York, 1976.

15. L. Michel, "Les brisures spontanees de symetrie en physique," Jour. de Physique 36 (1975), C7-41.

16. H. Othmer, "Spatial and temporal pattern formation," Annals, New York Academy of Sciences 316 (1979).

17. P.H. Roberts, "Dynamo Theory," in Mathematical Problems in the Geophysical Sciences, Lectures in Applied Mathematics, vol. XIV, American Math. Society, Providence, 1971.

18. D.H. Sattinger, "Topics in Stability and Bifurcation Theory," Lecture Notes 309, Springer-Verlag, New York, 1973.

19. A.M. Turing, "The Chemical Basis of Morphogenesis," Philo. Trans. Roy. Soc. London B237 (1952), 37-72.

20. H.R. Wilson, J.D. Cowan, "Excitatory and Inhibitory Interactions in Localized Populations of Model Neurons," Biophysical Journal 12 (1972), 1-24.

II

MATHEMATICAL PRELIMINARIES

In this chapter we outline the fundamental mathematical tools which are necessary for the analysis of bifurcation problems. Since the proofs of the basic results are readily available we shall not dwell on them here. Instead we shall concentrate on the concepts and statements of the theorems. We assume the reader is already familiar with such notions as Hilbert space, Banach space, and linear operators, as well as some simple facts about partial differential equations.

1. Linear Functional Analysis

A Banach space is a complete normed linear vector space. The norm $\|\ \|$ satisfies $\|\lambda u\| = |\lambda|\,\|u\|$ and $\|u+v\| \le \|u\| + \|v\|$; also $\|u\| = 0$ only if $u = 0$. A linear operator from a Banach space $\mathcal{E}$ to $\mathcal{F}$ is continuous if and only if it is bounded: $\|Lu\|_{\mathcal{F}} \le M\|u\|_{\mathcal{E}}$. An operator L is an isomorphism from $\mathcal{E}$ to $\mathcal{F}$ if it defines a bicontinuous one-to-one relationship between $\mathcal{E}$ and $\mathcal{F}$. That is, L and L^{-1} are both continuous. It is a consequence of the Open Mapping Theorem of basic functional analysis that L^{-1} is continuous if L maps $\mathcal{E}$ onto $\mathcal{F}$ in a one-to-one, continuous manner. (The open mapping theorem states that if $L\colon \mathcal{E} \to \mathcal{F}$ is one-to-one, bounded, and surjective then L takes open sets to open sets.) The fact that L is an isomorphism from $\mathcal{E}$ to $\mathcal{F}$ can be expressed by the inequality

$$C_1\|u\|_{\mathcal{E}} \leq \|Lu\|_{\mathcal{F}} \leq C_2\|u\|_{\mathcal{E}} \quad . \tag{2.1}$$

The kernel and range of L are respectively the subspaces

$$\mathcal{N} = \{u : Lu = 0\} \;, \quad \mathcal{R} = \{f : Lu = f \text{ for some } u \in \mathcal{E}\}$$

If $\mathcal{N}$ and $\mathcal{R}$ are closed subspaces, $\mathcal{N}$ of finite dimension and $\mathcal{R}$ of finite codimension, then L is said to be a Fredholm operator. The index of L in this case is $\nu = \dim \mathcal{R} - \dim \mathcal{N}$. In these notes we shall always assume $\nu = 0$. In bifurcation theory, one must consider non-invertible operators ; they are not always Fredholm operators (for example when small divisor problems arise as in celestial mechanics), but in these notes we shall confine ourselves to those problems in which the operators in question are Fredholm.

In order to put the remarks below in perspective, let us recall a basic construction in elementary group theory. A homomorphism ρ from a group G_1 to a group G_2 is a mapping which preserves group structure: $\rho(ab) = \rho(a)\rho(b)$. The kernel $\mathcal{N}$ of ρ is the set of elements in G_1 which are mapped into the identity element in G_2. Both $\mathcal{N}$ and $\rho(G_1)$ are subgroups. The first homomorphism theorem of group theory states that $\mathcal{N}$ is a normal subgroup of G_1 and that ρ lifts to an isomorphism between the factor group $G_1/\mathcal{N}$ and the group $\rho(G_1)$.

We can view a linear operator L mapping a vector space $\mathcal{E}$ to $\mathcal{F}$ as a homomorphism from one group to another. Given a subspace $\mathcal{N}$ of $\mathcal{E}$ the cosets of $\mathcal{N}$ are the sets $u + \mathcal{N}$, $u \in \mathcal{E}$. The quotient space $\mathcal{E}/\mathcal{N}$ is the vector space consisting of cosets and this is a Banach space with the norm

$$\|u + \mathcal{N}\| = \inf_{x \in \mathcal{N}} \|u + x\| .$$

Algebraically we may view $\mathcal{E}$ and $\mathcal{F}$ as abelian groups (with vector addition the group operation) and L as a homomorphism from $\mathcal{E}$ to $\mathcal{F}$. The kernel of L is then the null space $\mathcal{N}$ and L becomes an isomorphism between the factor group $\mathcal{E}/\mathcal{N}$ and the range space $\mathcal{R} \subset \mathcal{F}$. This observation is fundamental to the Lyapounov-Schmidt procedure in the reduction of bifurcation problems; we shall come back to a more concrete version of it below.

A projection is a bounded linear operator P such that $P^2 = P$. The operator $Q = I - P$ is then also a projection. Given a vector space W and a direct sum decomposition $W = U + V$ where $U \cap V = \{0\}$ we may define the projection $Pu = x$ where $u = x + y$, $x \in U$ and $y \in V$. In applications to bifurcation theory the projections onto the kernel and range of an operator L play a fundamental role. If $L: \mathcal{E} \to \mathcal{F}$ and $\mathcal{E} \subset \mathcal{F}$ then it makes sense to talk of projections P which commute with L: $LP = PL$. Such a projection can be constructed as follows.

Let $\mathcal{N} = [\varphi_1, \ldots, \varphi_n]$. Any projection P onto $\mathcal{N}$ must then take the form

$$Pu = \sum_{j=1}^{n} c_j(u)\varphi_j ,$$

where the $c_j(u)$ are complex coefficients depending on u. Clearly, these c_j must be linear functionals and so there are elements $\varphi_1^*, \ldots, \varphi_n^*$ in $\mathcal{E}^*$ such that $c_j(u) = \langle u, \varphi_j^* \rangle$. Now, if we require that $P^2 = P$ we see that we must have $\langle \varphi_r, \varphi_s^* \rangle = \delta_{rs}$ (for example use the fact that $P\varphi_j = \varphi_j$). Now use the requirement that PL = LP and we get

$$\sum_{j=1}^{n} \langle Lu, \varphi_j^* \rangle \varphi_j = 0$$

for all u; hence $L^* \varphi_j^* = 0$. That is, the φ_j^* must be null functions for the adjoint operator.

<u>The Fredholm Alternative</u>. Let $L: \mathcal{E} \to \mathcal{F}$ be a bounded linear operator. The adjoint operator L^* is then a continuous mapping from $\mathcal{F}^*$ to $\mathcal{E}^*$. Suppose that L has null functions $\varphi_1, \ldots, \varphi_n$ and that $\mathcal{R}(L)$ has co-dimension n. We describe $\mathcal{R}(L)$ by

$$\mathcal{R}(L) = \{ f \mid \langle f, \psi_j^* \rangle = 0, \ j = 1, \ldots, n\}$$

where $\psi_j^* \in \mathcal{F}^*$. Clearly, we have $\langle Lu, \psi_j^* \rangle = 0$ for all u, which implies that $\langle u, L^* \psi_j^* \rangle = 0$. The Fredholm alternative holds for an operator L if Lu = f is solvable whenever f is annihilated by all null vectors of L^* in the dual space. It will always be assumed in these notes that the Fredholm alternative is valid.

Let G be a group and $\mathcal{E}$ a vector space. A <u>representation</u> of G on $\mathcal{E}$ is a homomorphism from G to the group of linear transformations on $\mathcal{E}$. Thus $T_{g_1} T_{g_2}$ and $T_e = I$ where e is the identity element of G and I is the identity transformation. For example, let $\mathcal{E}$ be the space of continuous functions on $\mathbb{R}^3$ and let G be the group of rigid motions. Define $(T_g u)(x) = u(g^{-1}x)$; this is a representation of the group of rigid motions. We shall come back to group representations in more detail later on.

2. Elliptic and Parabolic Partial Differential Equations

The most important Banach spaces in applications to partial differential equations are the Holder and Sobolev spaces. If Ω is an open domain in $\mathbb{R}^n$ the Holder space $C_\alpha(\Omega)$ $(0 < \alpha \leq 1)$ is the class of continuous functions u on $\overline{\Omega}$ for which

$$H_\alpha(u) = \sup_{x, y \in \overline{\Omega}} \frac{|u(x) - u(y)|}{|x - y|^\alpha} < +\infty.$$

The norm on C_α is

$$\|u\|_\alpha = \sup_{x \in \overline{\Omega}} |u(x)| + H_\alpha(u) ,$$

and C_α is complete under this norm. More generally, it is important to consider the related spaces $C_{k,\alpha}(\Omega)$ consisting of functions whose partial derivatives up to order k have finite Holder norms. The norm $\| \ \|_{k,\alpha}$

is computed by summing the Hölder norms of u and all its derivatives up to k. Vector-valued functions can be treated in a similar manner: one simply sums over all the components of u. We shall need to consider spaces of vector-valued functions when we come to systems of partial differential equations.

The Sobolev spaces $W_{k,p}(\Omega)$ consist of functions u which have weak derivatives up to order k lying in $L_p(\Omega)$. The function u_i is the weak derivative of w with respect to x_i if

$$\int_\Omega u_i \varphi \, d\underline{x} = -\int_\Omega w \frac{\partial \varphi}{\partial x_i} \, d\underline{x}$$

for all $\varphi \in C_0^\infty(\Omega)$. We use the multi-index notation

$$\alpha = (\alpha_1, \dots, \alpha_n) \ , \quad |\alpha| = \alpha_1 + \dots + \alpha_n$$

$$D^\alpha w = \frac{\partial^{|\alpha|} w}{\partial x_1^{\alpha_1} \cdots \partial x_n^{\alpha_n}}$$

where $\alpha_1, \dots, \alpha_n$ are non-negative integers. If $D^\alpha w$ denotes the weak derivative for w then the norm for $W_{k,p}(\Omega)$ is

$$\|u\|_{k,p} = \sum_{|\alpha| \leq k} \|D^\alpha u\|_p \ .$$

The Hölder spaces have an especially nice property, namely they are actually Banach algebras; that is, the norm $\| \ \|_{k,\alpha}$ satisfies

$$\|uv\|_{k,\alpha} \leq C \|u\|_{k,\alpha} \|v\|_{k,\alpha} \tag{2.2}$$

for some positive constant C. By suitably renormalizing the norm $\| \ \|_{k,\alpha}$ (that is, by multiplying it by C^{-1}) we obtain the inequality

$$\|uv\|_{k,\alpha} \leq \|u\|_{k,\alpha}\|v\|_{k,\alpha}$$

making $C_{k,\alpha}$ a Banach algebra. Such an inequality is especially useful in dealing with nonlinear problems. For example, the nonlinear term $N(u) = (u\cdot\nabla)u$ in the Navier-Stokes equations is $u_j \frac{\partial u_i}{\partial x_j}$ and satisfies the inequality

$$\|N(u)\|_{0,\alpha} = \|u_j \frac{\partial u_i}{\partial x_j}\|_{0,\alpha} \leq \|u\|_{0,\alpha}\|u\|_{1,\alpha} \leq \|u\|^2_{1,\alpha} \tag{2.3}$$

(For vector-valued functions we simply sum the Hölder norms over the components.) This inequality means that the mapping $u \to N(u)$ is a continuous mapping from $C_{1,\alpha}$ to $C_{0,\alpha}$. (The proof is left to the reader; the mapping is even Lipschitz continuous locally.)

On the other hand, the mapping $u \to u^2$ is not continuous from $L_2(\Omega)$ to $L_2(\Omega)$. In fact, it is not even defined everywhere. For this reason L_2 spaces are sometimes (not always) inappropriate for treating nonlinear problems. The spaces $W_{k,p}$ are Banach algebras if p is large enough; specifically, p must satisfy the inequality $p > n/m$. (This is proved in R. Adams [1], p. 115 by expanding $D^\alpha[uv]$ according to Leibniz's rule for the differentiation of products and estimating each term.

The Holder spaces, as well as the Sobolev spaces and their more worldly cousins, the interpolation spaces, fit in very nicely with the theory of elliptic partial differential equations. Let me first state the results for the Laplace operator and follow that with a general discussion of the modern theory of elliptic operators according to Agmon, Douglis, and Nirenberg [2].

We consider the linear operator L given by

$$Lu = (\Delta u, u|_{\partial\Omega})$$

defined on the class $C_{2,\alpha}(\Omega)$. We assume Ω is a bounded domain in $\mathbb{R}^n$ with smooth $(C_{2,\alpha})$ boundary $\partial\Omega$. The operation $u \to u|_{\partial\Omega}$ is a trace or boundary operator. It takes functions defined on $\bar{\Omega}$ and maps them (by restriction) into functions defined on $\partial\Omega$. More generally, we could consider boundary operators of the form

$$Bu = \frac{\partial u}{\partial \tau} + gu$$

where $\frac{\partial}{\partial \tau}$ is a non-tangential directional derivative of u at the boundary and g is a smooth function defined on $\partial\Omega$.

In any event, the operator L is a linear operator from $C_{2,\alpha}(\Omega)$ to the Cartesian product $C_\alpha(\Omega) \times C_{2,\alpha}(\partial\Omega)$. Denoting by $\mathfrak{E}$ the Banach space $C_{2,\alpha}(\Omega)$ and by $\mathfrak{F}$ the product space $C_\alpha(\Omega) \times C_{2,\alpha}(\partial\Omega)$, we see that L is a bounded mapping from $\mathfrak{E}$ to $\mathfrak{F}$.

Furthermore, L is one-to-one, as follows from the uniqueness theory for the solution of the boundary value problem $Lu = f$. That is, if $f = (f_1, f_2)$ where $f_1 \in C_\alpha(\Omega)$ and $f_2 \in C_{2,\alpha}(\partial\Omega)$, then the boundary value problem

$$\Delta u = f_1 \qquad u|_{\partial D} = f_2 \tag{2.4}$$

has at most one solution. (One way to prove this is by applying the maximum principle for second order elliptic operators.) Not only that, but the boundary value problem (2.2) is always solvable; this follows from classical methods in potential theory. One may derive by analytic methods (for example, by analyzing an integral representation of the solution of (2.4)) the a priori estimate

$$\|u\|_{2,\alpha} \leq C\|f\| = C(\|f_1\|_\alpha + \|f_2\|_{2,\alpha}^{\partial\Omega}) \tag{2.5}$$

for some constant C independent of f. As we saw above, however, the bounded invertibility of L follows from the fact that L is continuous, one-to-one, and onto; so (2.5) is really a consequence of the open mapping theorem of functional analysis.

The a priori inequalities above, that is, the fact that L defines an isomorphism between the domain and range spaces, holds more generally for elliptic systems of partial differential equations. We include here a brief description of the modern theory (see Agmon, Douglis, and Nirenberg [2]).

Consider a system of partial differential equations

$$\sum_{j=1}^{N} \ell_{ij}(x, D)u_j(x) = f_i(x) \tag{2.6}$$

where $D = (\frac{\partial}{\partial x_1}, \ldots, \frac{\partial}{\partial x_n})$. Let $\xi = (\xi_1, \ldots, \xi_n)$ and consider the polynomials $\ell_{ij}(x, \xi)$ in ξ. Construct a system of integer weights

$s_1, \ldots, s_N, t_1, \ldots, t_N$ such that

$$\deg \ell_{ij}(x, \xi) \leq s_i + t_j \qquad i, j = 1, \ldots, N ;$$

$\ell_{ij} = 0$ if $s_i + t_j < 0$. The system (2.6) is elliptic if

$$L(x, \xi) = \det(\ell'_{ij}(x, \xi)) \neq 0 \text{ for real } \xi \neq 0 ,$$

where ℓ'_{ij} consists only of those terms of ℓ_{ij} which are of order $s_i + t_j$.

Let the elliptic system above be supplemented by boundary conditions

$$\sum_{j=1}^{N} B_{hj}(x, D)u_j(x) = \phi_h(x) \quad \text{on} \quad \Gamma$$

where Γ is the boundary of the domain and the B_{hj} are boundary operators. Again there are systems of integer weights $r_1, \ldots, r_m$ and $t_1, \ldots, t_N$ as before such that

$$\deg B_{hj}(x, \xi) \leq r_h + t_j .$$

The principle parts B'_{hj} are those terms in B_{hj} for which equality holds above. The boundary conditions must complement the differential equations in the correct manner in order for coercive inequalities (the analogue of (2.5)) to hold. The necessary and sufficient complementing conditions can be formulated as an algebraic condition, as follows.

At any point P of Γ let $\hat{n}$ denote the normal and $\xi \neq 0$ any tangent vector to Γ. Denote by $\tau_k^+(x, \xi)$, $k = 1, \ldots, m$ the m roots in τ with positive imaginary part of the characteristic equation $L(x, \xi + \tau\hat{n}) = 0$. (It can be shown that all elliptic systems in three or more independent variables satisfy the following supplementary condition:

$L(x, \xi)$ is of even degree $2m$ and for every pair of independent real vectors ξ and ξ', the polynomial $L(x, \xi + \tau\xi')$ in the complex variable τ has precisely m roots with positive imaginary part.) Accordingly, set

$$M^{+}(x, \xi, \tau) = \prod_{h=1}^{m} (\tau - \tau_h^{+}(x, \xi)).$$

Let $L^{jk}(x, \xi + \tau\hat{n})$ be the matrix adjoint to $\ell'_{ij}(x, \xi + \tau\hat{n})$. The necessary and sufficient condition for the boundary value problem to be coercive is the

Complementing Boundary Condition. For $x \in \Gamma$ and any real non-zero vector ξ tangent to Γ at x consider the matrix elements

$$\sum_{j=1}^{N} B'_{hj}(x, \xi + \tau\hat{n}) L^{jk}(x, \xi + \tau\hat{n})$$

as polynomials in τ. We require the rows of this matrix to be linearly independent modulo $M^{+}(x, \xi, \tau)$, i.e.,

$$\sum_{h=1}^{m} C_h \sum_{j=1}^{N} B'_{hj} L^{jk} \equiv 0 \quad (\mathrm{mod}\ M^{+})$$

implies the constants C_h are all zero. (That is, if the left-hand side above is a constant multiple of $M^{+}(\tau)$, then the coefficients C_h are all zero.)

Roughly speaking, when one has an elliptic system (in the above sense) on a smoothly bounded finite domain in $\mathbb{R}^n$ and the complementing conditions hold, then the elliptic system defines an isomorphism (bicontinuous) whenever it is one-to-one.

For example, consider the Stokes system of equations which arises in fluid mechanics

$$\begin{aligned} \Delta u_k - \frac{\partial p}{\partial x_k} &= f_k \qquad k = 1,2,3 \\ \frac{\partial u_i}{\partial x_i} &= f_4 \end{aligned} \tag{2.7}$$

with boundary conditions

$$u_i|_\Gamma = g_i \ .$$

P. Fife [3] has shown that this is an elliptic system in the sense described above and that the complementing condition holds. The integer weights are $s_1 = s_2 = s_3 = 0$, $s_4 = -1$ and $t_1 = t_2 = t_3 = 2$, $t_4 = 1$; and $L(\xi) = |\xi|^6$, so the system is elliptic. Let us check the complementing condition on a plane whose normal lies along the z-axis. A tangent vector ξ has the form $\xi = \xi_1\hat{i} + \xi_2\hat{j}$ and the normal vector is $\hat{k}$. $L(\xi_1\hat{i} + \xi_2\hat{j} + \tau\hat{k}) = (\xi_1^2 + \xi_2^2 + \tau^2)^3$. The positive imaginary roots of this equation are $\tau = i\rho$, where $\rho^2 = \xi_1^2 + \xi_2^2$, as $M^+(\xi,\tau) = (\tau - i\rho)^3$. Letting L^{ij} be the matrix adjoint to ℓ_{ij}, a computation shows that [2]

$$\sum_{j=1}^{4} B_{hj}(\xi + \tau\hat{n})L^{jk}(\xi + \tau\hat{n}) =$$

$$-(\rho^2 + \tau^2)\begin{bmatrix} (\xi_2^2 + \tau^2) & -\xi_1\xi_2 & -\xi_1\tau & -\xi_1(\rho^2+\tau^2) \\ -\xi_1\xi_2 & \xi_1^2 + \tau^2 & -\xi_2\tau & -\xi_2(\rho^2+\tau^2) \\ -\xi_1\tau & -\xi_2\tau & \rho^2 & 3i\rho\tau^2 + 2\rho\tau - i\rho^3 \end{bmatrix}.$$

For any $\rho \neq 0$, this matrix has rank 3 when its elements are considered as polynomials in τ; thus its rows are linearly independent.

As a consequence of the elliptic nature of (2. 7) one obtains the following a priori estimates for its solutions

$$|u|_{2,\alpha} + |p|_{1,\alpha} \le C|f|_\alpha$$

on a bounded domain Ω. The Hölder norms here refer to the vector-valued components u and f; p is a scalar function (the hydrodynamic pressure).

Estimates for Hölder norms of solutions of elliptic systems are referred to as Schauder estimates since these were first derived by Schauder in the 1930's for general second order elliptic equations. Similar types of estimates are obtainable for the Sobolev $W^{k,p}$ norms. Moreover, there are analogous estimates for parabolic systems. For example, when dealing with parabolic (time evolution) equations which are of first order in the time variable, one introduces the norms on the time cylinder $D = \Omega \times [0,T]$:

$$\| u \|_{2\alpha,\alpha} = \sup_{(x,t)\in D} |u(x,t)| + \sup_{\substack{(x,y)\in\Omega \\ s,t\in[0,T]}} \frac{|u(x,s) - y(y,t)|}{|x-y|^{2\alpha} + |s-t|^{\alpha}} ,$$

where $0 < \alpha \leq \frac{1}{2}$. Then, for example, for the linear time dependent Stokes equations, one obtains the following a priori estimates

Theorem (Solonnikov). Let u_i, p satisfy the initial value problem

$$\frac{\partial u_i}{\partial t} - \Delta u_i = -\frac{\partial p}{\partial x_i} + f_i$$

$$\frac{\partial u_i}{\partial x_i} = 0 \quad , \quad u_i(x,0) = \psi_i(x) \quad , \quad u_i|_{\partial D} = 0 .$$

Let $\operatorname{div} \psi = 0$, $\psi \in C_{2,2\alpha}(\overline{D})$, $f \in C_{2\alpha,\alpha}(\overline{D}_T)$, $(\overline{D}_T = D \times [0,T])$, and let $\partial D \in C_{2,2\alpha}$. Then $u \in C_{2,2\alpha;1,\alpha}$ and there is a constant C such that

$$(\| u \|_{2,2\alpha;1,\alpha} + \| p \|_{1,2\alpha;1,\alpha}) \leq C(\| f \|_{2\alpha,\alpha} + \| \psi \|_{2,2\alpha}).$$

This a priori result was derived by Solonnikov using hydrodynamic potentials. A discussion of these results is contained in Ladyzhenskaya's book [5].

The result described above is important in deriving the basic analytical results which are necessary in extending Hopf's bifurcation theorem (of periodic solutions) to parabolic systems of partial differential equations.

3. Nonlinear Functional Analysis

Let $B(u, v)$ be a bilinear operator on a Banach space $\mathcal{E}$; we can obviously construct from B the quadratic operator $Q(u) = B(u, u)$. Conversely, given a quadratic operator Q — that is, one for which $Q(\lambda u) = \lambda^2 Q(u)$ — we can recover Q from a bilinear operator B. In fact, we can take B to be

$$B(u, v) = \frac{\partial^2}{\partial s \partial t} Q(su + tv)\Big|_{s=0, t=0} .$$

Then B is linear in u and v and $Q(u) = \frac{1}{2} B(u,u)$. These observations can be generalized immediately to operators of any degree (see [6], p. 67). Some simple quadratic operators are $u \to u^2$ where u is a scalar valued function on a domain Ω, and $Q(u) = u_j \frac{\partial u_i}{\partial x_j}$ where u_i is a vector-valued function on a domain Ω. The latter operator occurs in hydrodynamics, u_i being the components of the velocity field.

Let $F(u)$ be a nonlinear operator on a Banach space $\mathcal{E}$. F is said to be Frechet differentiable at a point u provided there is a linear operator A such that the quantity

$$R(u; h) = F(u+h) - F(u) - Au$$

is $o(h)$ as $h \to 0$; that is,

$$\lim_{\|h\| \to 0} \frac{\| R(u; h)\|}{\|h\|} = 0.$$

We shall denote the Frechet derivative of F by $F'(u)$. If F is Frechet differentiable at u then its Frechet derivative may be computed by the simple formula

$$F'(u)h = \lim_{t \to 0} \frac{F(u+th) - F(u)}{t} .$$

For example, the Frechet derivative of the operation u^2 is the linear multiplication operator $h \to 2uh$.

The usual theorems for differentiation of ordinary functions carry over to functional differentiation. The chain rule, for example, is still valid; its proof goes like this

$$\begin{aligned} F(G(t_0+h)) &= F(G(t_0) + G'(t_0)h + R_1(t_0; h)) \\ &= F(G(t_0)) + F'(G(t_0))[G'(t_0)h + R(t_0; h)] \\ &\qquad + R_2(G(t_0); G'(t_0)h + R_1(t_0; h)) \\ &= F(G(t_0)) + F'(G(t_0))G'(t_0)h + o(h) . \end{aligned}$$

Thus the Frechet derivative of the composite map is $F'(G(t_0))G'(t_0)$.

If F is a differentiable mapping from $\mathcal{E}$ to $\mathcal{F}$ its Frechet derivative $F'(u)$ is a bounded linear mapping from $\mathcal{E}$ to $\mathcal{F}$. The space of bounded linear mappings from $\mathcal{E}$ to $\mathcal{F}$ is itself a Banach space, $\mathcal{L}(\mathcal{E}, \mathcal{F})$; and if we regard u as a variable we again have a nonlinear mapping, this time from $\mathcal{E}$ to $\mathcal{L}(\mathcal{E}, \mathcal{F})$. Thus it makes sense to consider the Frechet derivative of $F'(u)$. This must be a linear mapping from $\mathcal{E}$ to $\mathcal{L}(\mathcal{E}, \mathcal{F})$. For fixed $u \in \mathcal{E}$ let us denote such a mapping by $v \to B_u(v)$, but since B_u is linear in u as well, we may as well write $B(u, v)$; and the space of linear operators $\mathcal{L}(\mathcal{E}, \mathcal{L}(\mathcal{E}, \mathcal{F}))$ may be identified with bilinear mappings from $\mathcal{E} \times \mathcal{E}$ to $\mathcal{F}$. One can proceed in this way and prove Taylor's theorem for the nonlinear operator F. We get

$$F(u+x) = F(u) + F'(u)x + B_2(u; x, x) + B_3(u; x, x, x) + \dots$$
$$+ B_n(u; x, \dots, x) + R_n(u, x)$$

where $\|R_n(u, x)\| = o(\|x\|^n)$. F is said to be n times Frechet differentiable if the above expansion holds, with each B_j satisfying the inequality

$$\|B_j(u; x_1, \dots, x_j)\| \leq M\|x_1\| \cdots \|x_j\|.$$

An operator $F: \mathcal{E} \to \mathcal{F}$ is strongly analytic in the neighborhood $\|x - x_0\| < \delta$ if there is a sequence of bounded homogeneous operators A_k of degree k such that the series

$$F(x_0 + v) = \sum_{k=0}^{\infty} A_k(v)$$

is convergent for $\|v\| < \delta$. By A_0 we understand a fixed vector in $\mathcal{F}$. F is weakly analytic if the complex-valued function $\langle F(zx), \varphi^* \rangle$ is is analytic in the usual sense, φ^* being an element of the dual space $\mathcal{F}^*$. One can prove that F is strongly analytic if and only if it is weakly analytic; and, furthermore, that F is strongly analytic if it is Frechet differentiable on a complex Banach space $\mathfrak{E}$. [5]

We come finally to the implicit function theorem, which will prove itself to be a fundamental tool in bifurcation theory.

Implicit Function Theorem: Let $G(\lambda, u)$ be a mapping from $\Lambda \times \mathfrak{E}$ to $\mathfrak{F}$, where Λ, $\mathfrak{E}$, and $\mathfrak{F}$ are Banach spaces; and suppose that $G(\lambda_0, u_0) = 0$, that G is continuously Frechet differentiable in a neighborhood of (λ_0, u_0) in $\Lambda \times \mathfrak{E}$, and that $G_u(\lambda_0, u_0)$ is an isomorphism from $\mathfrak{E}$ to $\mathfrak{F}$. Then for sufficiently small $|\lambda - \lambda_0|$ there exists a function $u(\lambda)$ such that $u(\lambda_0) = u_0$ and $G(\lambda, u(\lambda)) \equiv 0$. The graph $(\lambda, u(\lambda))$ is the locally (in $\Lambda \times \mathfrak{E}$) unique zero set of G. The function $u(\lambda)$ is C^k (k times continuously differentiable) if G is C^k; and u is analytic in λ if G is analytic — in particular, if Λ, $\mathfrak{E}$, and $\mathfrak{F}$ are complex Banach spaces, then $u(\lambda)$ is an analytic function of λ.

The proof of the implicit function theorem in a Banach space goes back to Graves and Hildebrandt (1923) [4] . The proof is based on a contraction mapping argument. (See also [6])

References

1. R. Adams, Sobolev Spaces, Academic Press, 1975.
2. S. Agmon, A. Douglis, L. Nirenberg, "Estimates near the boundary for solutions of elliptic partial differential equations satisfying general boundary conditions," II Comm. Pure Appl. Math. 17 (1964), 35-92.
3. P. Fife, "The Benard problem for general fluid dynamical equations and remarks on the Boussinesq approximation," Indiana University Mathematics Journal, 20 (1970), 303-326.
4. L. M. Graves and T. H. Hildebrandt, "Implicit functions and their differentials in general analysis," Trans. Amer. Math. Soc. 29 (1923), 127-153.
5. O. A. Ladyzhenskaya, The Mathematical Theory of Viscous Incompressible Flow, 2nd English ed. Gordon and Breach, 1969.
6. D. H. Sattinger, Topics in Stability and Bifurcation Theory, Springer Mathematical Lecture Notes No. 309, 1973.
7. J. T. Schwartz, Nonlinear Functional Analysis, Gordon and Breach, New York, 1969.

III

STABILITY AND BIFURCATION

1. Principle of Linearized Stability

Consider an equation of the form $G(\lambda, u) = 0$ in the neighborhood of a known solution (λ_0, u_0). If $G_u(\lambda_0, u_0)$ is invertible then by the implicit function theorem there is a smooth curve $u = u(\lambda)$ of solutions through the point (λ_0, u_0). Bifurcation may occur, however, when $G_u(\lambda_0, u_0)$ is not invertible — for example, when it has a non-trivial kernel and a closed range of finite codimension. Now suppose that the time dependent equations governing the dynamics of the system take the form

$$\frac{\partial u}{\partial t} = G(\lambda, u). \tag{3.1}$$

Consider the initial value problem (3.1) on a Banach space $\mathcal{E}$. An equilibrium solution u_0 is stable if given any $\epsilon > 0$ there is a $\delta > 0$ such that whenever $\|u(0) - u_0\| < \delta$, $\|u(t) - u_0\| < \epsilon$ for all $t > 0$. Furthermore, u_0 is asymptotically stable if in addition $u(t) \to u_0$ as $t \to \infty$. If (3.1) is a system of ordinary differential equations (that is, $\mathcal{E} = \mathbb{R}^n$) then Lyapounov's first theorem states that an equilibrium solution u_0 is asymptotically stable if all the eigenvalues of the Frechet derivative $G_u(\lambda, u_0)$ have negative real parts; and u_0 is unstable if some eigenvalues of $G_u(\lambda, u_0)$ have positive real part. Lyapounov's theorem has been extended to nonlinear systems of partial differential equations which are "dissipative" — for example, to the Navier-Stokes equations as well as to general parabolic systems. We shall assume the validity of the result in the present discussion; of course, it must be checked in each particular case. In the case of hydrodynamic stability, the reader may consult the treatise by D. Joseph [5] for a thorough bibliography. For now, we state the

PRINCIPLE OF LINEARIZED STABILITY: An equilibrium solution u_0 of (3.1) is stable if the spectrum of the linear operator $G_u(\lambda, u_0)$ is contained strictly in the left half plane; it is unstable if part of the spectrum of $G_u(\lambda, u_0)$ lies in the right half plane.

Now suppose we have a known solution $u(\lambda)$ of (3.1) — that is, suppose we can write down an explicit solution for all values of λ. This is often the case in hydrodynamical problems, for example. We linearize (3.1) about $u(\lambda)$, obtaining a linear operator

$$L(\lambda) = G_u(\lambda, u(\lambda)).$$

As λ varies, the spectrum of $L(\lambda)$ also varies. Suppose that as λ crosses a critical value λ_0 one or more eigenvalues of $L(\lambda)$ cross the imaginary axis from the left to the right half plane. This is precisely the situation where $u(\lambda)$ loses stability as λ crosses λ_0, and at this point some kind of bifurcation may take place. We begin by discussing the case in which a simple eigenvalue crosses through the origin as λ crosses λ_0.

2. Bifurcation of a Simple Eigenvalue

Theorem 3.1. Let $G(\lambda, u)$ be an analytic mapping and suppose there is a known solution $u(\lambda)$ of (3.1) which becomes unstable by virtue of a simple isolated eigenvalue $\sigma(\lambda)$ of $G_u(\lambda, u)$ crossing the origin as λ crosses λ_0. Assume $\sigma(\lambda_0) = 0$ and $\sigma'(\lambda_0) > 0$. Then there is a smooth nontrivial solution curve $\lambda(\epsilon)$, $u(\epsilon)$ bifurcating from the given solution $u(\lambda)$ at (λ_0, u_0). The bifurcating solutions are stable when they appear above criticality and unstable when they occur below criticality.

The three possible situations are depicted below in Figure 3.1.

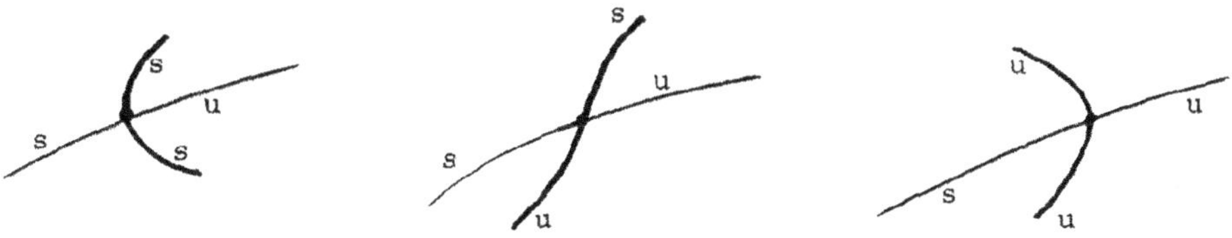

(a) (b) (c)

Figure 3.1

Theorem 3.1 was first stated by E. Hopf in his 1942 article for ordinary differential equations, but there were certain restrictive conditions on the function $\lambda(\epsilon)$. Namely, if $\lambda(\epsilon) = \lambda_1 \epsilon + \lambda_2 \epsilon^2 + \dots$ then Hopf required $\lambda_1 \neq 0$ or if $\lambda_1 = 0$ then $\lambda_2 \neq 0$. Hopf's method was to compute the perturbation series for the critical eigenvalue along the bifurcating branch. This computation is carried out in section 4. The stability results of Theorem 3.1 were generalized to infinite dimensional problems by using topological degree theory [9]; there the conditions on λ_1 and λ_2 were not needed.

A complete proof of the full results, by analytical methods, will be given in the next chapter.

<u>Proof.</u> We can always assume that $G(\lambda, 0) = 0$; otherwise replace G by $G(\lambda, u(\lambda) + u) - G(\lambda, u(\lambda))$. Similarly, we can take $\lambda_0 = 0$. We shall furthermore assume that $G(0, u) \neq 0$ for small $u \neq 0$.

Let $L(\lambda) = G_u(\lambda, 0)$ and $L_0 = G_u(0, 0)$. By assumption L_0 has a simple isolated eigenvalue at the origin and so has a one-dimensional kernel and a closed range of codimension one. Assume G is a mapping from $\Lambda \times \mathcal{E}$ to $\mathcal{F}$, so that L_0 is a bounded linear mapping from $\mathcal{E}$ to $\mathcal{F}$. Let $\mathcal{N} = \ker L_0 = [\varphi_0]$; and let the range of L_0 be characterized by $\mathcal{R} = \{ f \mid \langle f, \varphi_0^* \rangle = 0$, where $\varphi_0^* \in \mathcal{F}^* \}$. By the Fredholm alternative $L_0 u = f$ is solvable if and only if $\langle f, \varphi_0^* \rangle = 0$. We assume that $\langle \varphi_0, \varphi_0^* \rangle \neq 0$ and normalize this quantity to be one.

We do not assume here that $\mathcal{E} \subset \mathcal{F}$. Let P_1 be a projection onto $\mathcal{N}$ and Q_2 a projection onto $\mathcal{R}$. Then we define $Q_1 = I - P_1$ and $P_2 = I - Q_2$. Decompose equation (3.1) as follows

$$u = \alpha\varphi_0 + \psi \,, \quad Q_1\psi = \psi \tag{3.2a}$$

$$Q_2 G(\lambda, \alpha\varphi_0 + \psi) = 0 \tag{3.2b}$$

$$P_2 G(\lambda, \alpha\varphi_0 + \psi) = 0 \tag{3.2c}$$

Since $Q_1\psi = \psi$ the first equation is equivalent to

$$H(\lambda, \alpha, \psi) \equiv Q_2 G(\lambda, \alpha\varphi_0 + Q_1\psi) = 0 \tag{3.3}$$

Now $H(\lambda_0, 0, 0) = 0$ and $H_\psi(\lambda_0, 0, 0) = Q_2 G_u(\lambda_0, 0) Q_1$. (The reader should check this.) H_ψ is a one-to-one mapping from $Q_1\mathcal{E}$ onto $\mathcal{R}$. In fact, if $H_\psi v = 0$ then $G_u(\lambda_0, 0)Q_1\psi = 0$ since Q_2 is a projection onto the range of $G_u(\lambda_0, 0)$; but then $Q_1\psi \in \mathcal{N}$ and therefore $\psi = 0$. Furthermore, H_ψ is surjective since $G_u(\lambda_0, 0)u = f$ is always solvable for $f \in \mathcal{R}$; and therefore the vector u also satisfies $Q_2 G_u(\lambda_0, 0) Q_1 u = f$. Since H_ψ is one-to-one and surjective, it is an isomorphism by the open mapping theorem, and therefore equation (3.3) is locally solvable by the implicit function theorem. Denote the solution by $\psi = \psi(\alpha, \lambda)$; it is an analytic function of α and λ as complex variables. Now take this function ψ and substitute it into equation (3.2c). The projection P_2 projects onto a one-dimensional subspace, so it is of the form $P_2 f = \langle f, \varphi_0^* \rangle \varphi_0$; and equation (3.2c) is equivalent to the scalar equation

$$F(\lambda, \alpha) = \langle G(\lambda, \alpha\varphi_0 + \psi), \varphi_0^* \rangle = 0. \tag{3.4}$$

Equation (3.4) is called the <u>bifurcation equation.</u> Solutions of (3.4) are in 1-1 correspondence with solutions of (3.1) in the neighborhood of the bifurcation point. Expand F in a power series in λ and α; since $F(\lambda, 0) \equiv 0$ by assumption ($u = 0$ is always a solution and this corresponds to $\alpha = 0$ in the bifurcation equations), each term of F contains α as a factor:

$$F(\alpha, \lambda) = \alpha(A_{10}\lambda + A_{20}\lambda^2 + \dots). \tag{3.5}$$

We compute the term A_{10} by computing $F_{\alpha\lambda}$. By the chain rule,

$$F_\alpha(\alpha, \lambda) = \langle G_u(\lambda, \alpha\varphi_0 + \psi)[\varphi_0 + \psi_\alpha], \varphi_0^* \rangle$$

$$F_\alpha(0, \lambda) = \langle G_u(\lambda, 0)[\varphi_0 + \psi_\alpha(0, \lambda)], \varphi_0^* \rangle$$

$$F_{\alpha\lambda}(0,0) = \langle G_{u\lambda}(0, 0)[\varphi_0 + \psi_\alpha(0, 0)] + G_u(0, 0)[\varphi_0 + \psi_{\alpha\lambda}(0,0)], \varphi_0^* \rangle$$

Now $\langle L_0 w, \varphi_0^* \rangle = 0$ for any w since by the Fredholm alternative φ_0^* is the adjoint null function, so the second term above is zero. (Recall that $L_0 = G_u(0, 0)$). As for the first term, $\psi_\alpha(0, 0) = 0$ as can be seen by differentiating (3.2b) with respect to α and setting $\alpha = \lambda = 0$. (Note first that $\psi(0, 0) = 0$.) Therefore

$$F_{\alpha\lambda}(0, 0) = \langle G_{u\lambda}(0, 0)\varphi_0, \varphi_0^* \rangle .$$

We now show that the quantity on the right is $\sigma'(0)$, where $\sigma(\lambda)$ is the critical eigenvalue. In fact, $L(\lambda) = G_u(\lambda, 0)$ and $L(\lambda)\varphi(\lambda) = \sigma(\lambda)\varphi(\lambda)$. Differentiating this expression with respect to λ and setting $\lambda = 0$

we obtain

$$L'(0)\varphi_0 + L_0\varphi'(\lambda) = \sigma'(0)\varphi_0 + \sigma(0)\varphi_0' = \sigma'(0)\varphi_0 .$$

But $L'(0) = G_{u\lambda}(0,0)$, and operating on this equation by the functional φ_0^* we obtain

$$< G_{u\lambda}(0,0)\varphi_0, \varphi_0^* > = \sigma'(0) < \varphi_0, \varphi_0^* > = \sigma'(0).$$

By hypothesis $\sigma'(0) \neq 0$ and so $F_{\alpha\lambda}(0,0) \neq 0$. This means the term $A_{10} \neq 0$; and so, after dividing out the α factor in $F(\lambda, \alpha)$ we are left with a function $g(\lambda, \alpha)$ for which $g_\lambda(0,0) \neq 0$. We apply the implicit function theorem to g, thereby obtaining a solution $\lambda = \lambda(\alpha)$ of $g(\lambda, \alpha) = 0$. This second solution represents the unique bifurcating solution branch.

We shall leave the proof of the stability results until the next chapter, where they will be derived as a consequence of general results on bifurcation and stability at multiple eigenvalues. From the assumption that $G(0,u)$ does not vanish for some interval $0 < \|u\| < \delta$ we may conclude that $F(\alpha, 0)$ does not vanish for small $\alpha \neq 0$. Therefore the Taylor series for F has the form

$$F(\alpha, \lambda) = \sigma'(0)\alpha\lambda + A_{k0}\alpha^k + \ldots \tag{3.6}$$

We shall use this expression in the analysis of the stability of the bifurcating solutions.

3. Power Series Expansions

We now discuss an alternative approach to the construction of the bifurcating solutions. We begin by treating the simple mathematical example

$$\Delta u + \lambda u + u^2 = 0 \qquad\qquad (3.7)$$
$$u\big|_{\partial D} = 0$$

The bifurcation points for this problem are the eigenvalues of the Laplacian. We can set the problem up as a mapping by defining

$$G(\lambda, u) = (\Delta u + \lambda u + u^2, u|_{\partial D})$$

as we did in Chapter II in discussing the isomorphism properties defined by elliptic boundary value problems.

Let (λ_0, φ_0) be an eigenpair of the Laplacian: $(\Delta + \lambda_0)\varphi_0 = 0$, and assume λ_0 is a simple eigenvalue. The projection P onto the subspace spanned by φ_0 is

$$Pu = \langle u, \varphi_0 \rangle \varphi_0$$

where

$$\langle u, \varphi_0 \rangle = \int_D u\varphi_0 \, dx \qquad\qquad (3.8)$$

We may still use the Hölder spaces $C_{2,a}$; then φ_0 is, strictly speaking, an element of the dual space, but considered as a linear functional it takes the form (3.8).

Now introduce a new parameter ϵ and write u and λ in the form

$$u = \epsilon\varphi_0 + \epsilon^2\psi \qquad\qquad (3.9)$$
$$\lambda = \lambda_0 + \epsilon\tau$$

where $P\psi = 0$ and $\psi = \psi(\epsilon)$, $\tau = \tau(\epsilon)$. Substituting the expansions (3.9) into (3.7) we get

$$(\Delta + \lambda_0)(\epsilon\varphi_0 + \epsilon^2\psi) + \epsilon^2\tau\varphi_0 + \epsilon^2\varphi_0^2 + O(\epsilon^3) = 0.$$

Dividing this equation by ϵ^2 we get

$$(\Delta + \lambda_0)\psi + \tau\varphi_0 + \varphi_0^2 + \epsilon H(\epsilon, \tau, \psi) = 0 \tag{3.10}$$

where H denotes general nonlinear terms in ψ and σ. Now decompose this equation by applying the projections P and Q; we get the system

$$\begin{aligned} \mathfrak{F}_1(\psi, \tau, \epsilon) &\equiv (\Delta + \lambda_0)\psi + Q[\varphi_0^2 + \epsilon H(\epsilon, \tau, \psi)] = 0 \\ \mathfrak{F}_2(\psi, \tau, \epsilon) &\equiv \tau + <\varphi_0^2, \varphi_0> + \epsilon <H(\epsilon, \tau, \psi), \varphi_0> = 0 \end{aligned} \tag{3.11}$$

The second equation is obtained by integrating (3.10) against φ_0 (We normalize φ_0 so that $<\varphi_0, \varphi_0> = 1$.) Now represent the system (3.11) as a single mapping $\mathfrak{F} = (\mathfrak{F}_1, \mathfrak{F}_2)$ with two components. $\mathfrak{F}$ is a Frechet differentiable mapping from $C_{2,\alpha} \times \mathbb{C}^2$ to $C_\alpha \times \mathbb{C}$. At $\epsilon = 0$ $\mathfrak{F}(\psi_0, \tau_0, 0) = 0$, where ψ_0 and τ_0 satisfy

$$(\Delta + \lambda_0)\psi_0 + Q\varphi_0^2 = 0$$

$$\tau_0 + <\varphi_0^2, \varphi_0> = 0$$

The Frechet derivative is the matrix operator

$$\frac{\partial(\mathfrak{F}_1, \mathfrak{F}_2)}{\partial(\psi, \tau)} = \begin{bmatrix} \Delta + \lambda_0 & 0 \\ 0 & 1 \end{bmatrix}$$

This operator is an isomorphism from $QC_{2,\alpha} \times \mathbb{C}$ to $QC_\alpha \times \mathbb{C}$. (The reader may check the invertibility for himself. The subspace QC_α is the set of all elements of C_α which satisfy the orthogonality condition $\langle f, \varphi_0 \rangle = 0$; but this is precisely the Fredholm alternative.) By the implicit function theorem, then, there is an analytic solution $\tau(\epsilon)$, $\psi(\epsilon)$ of the system (3.11).

Knowing that u and λ can be developed in a convergent power series, let us write

$$\lambda = \lambda_0 + \epsilon\lambda_1 + \dots$$

$$u = \epsilon\varphi_0 + \epsilon^2(\psi_0 + \epsilon\psi_1 + \dots).$$

Substituting these power series into (3.7) and separating out powers of ϵ we get

$$(\Delta + \lambda_0)\psi_0 + \lambda_1\varphi_0 + \varphi_0^2 = 0$$

$$(\Delta + \lambda_0)\psi_1 + \lambda_2\varphi_0 + \dots = 0$$

$$\vdots$$

$$(\Delta + \lambda_0)\psi_n + \lambda_{n+1}\varphi_0 + \dots = 0$$

where the dots in each case denote terms depending on the previous ψ_j's and λ_j's. In order to solve the first equation for ψ_0 we must choose λ_1 so that $\lambda_1 + \langle \varphi_0^2, \varphi_0 \rangle = 0$. Then the Fredholm alternative is satisfied; and we solve uniquely for ψ_0 by requiring $\langle \psi_0, \varphi_0 \rangle = 0$. We proceed in the same way at each step. Choose λ_{n+1} so that the Fredholm alternative is satisfied and then solve for ψ_n.

The same procedure described above for the simple problem (3.7) works in the more general case $G(\lambda, u) = 0$ provided G is analytic. Let us outline the proof. First we recall that we are assuming the Hopf condition $\sigma'(\lambda_0) > 0$. We expand $G(\lambda, u)$ in a Taylor series

$$G(\lambda, u) = G_u(\lambda, 0)u + R(\lambda, u)$$

where $R(\lambda, u)$ begins with quadratic terms in u. Now write $L(\lambda) = G_u(\lambda, 0) = L_0 + \lambda L_1 + \ldots$, and make the same substitutions (3.9) for u and λ as we did before. Applying the projections P and Q, we get ($\lambda_0 = 0$ here)

$$\epsilon^2(L_0\psi + \tau Q L_1\varphi_0 + QR_2(0, \varphi_0)) + O(\epsilon^3) = 0$$

$$\tau < L_1\varphi_0, \varphi_0^* > + < R_2(0, \varphi_0), \varphi_0 > + O(\epsilon) = 0$$

where R_2 denotes the quadratic terms in u of $R(\lambda, u)$.

Again dividing out by ϵ^2 we arrive at

$$L_0\psi + \tau Q L_1\varphi_0 + QR_2(0, \varphi_0) + Q\,O(\epsilon) = 0$$

$$\tau < L_1\varphi_0, \varphi_0^* > + < R_2(0, \varphi_0), \varphi_0^* > + O(\epsilon) = 0$$

where the expressions $O(\epsilon)$ indicate nonlinear terms in ψ, τ and ϵ which vanish at least to order ϵ. This new system can be solved via the implicit function theorem just as before.

We can arrive at a power series expansion also by treating the bifurcation equations $F(\alpha, \lambda) = 0$. Assuming F has the form (3.6), put $\alpha = \epsilon\xi$ and $\lambda = b\epsilon^k$, where b is a constant to be determined. Then

$$F(\epsilon\xi, b\epsilon^k) = \epsilon^{k+1}\xi(b\sigma'(0) + A_{k0}\xi^k) + O(\xi\epsilon^{k+2}).$$

Dividing by ϵ^{k+1}, we arrive at the equation

$$g(\epsilon, b, \xi) = b\sigma'(0) + A_{k0}\xi^k + O(\xi\epsilon).$$

When $\epsilon = 0$ we get the equation

$$b\sigma'(0) + A_{k0}\xi^k = 0 . \tag{3.12}$$

There are two cases to consider. i) k even; choose $b = \pm 1$ so that $b\sigma'(0)/A_{k0}$ is negative. Then for ξ take the positive root of (3.12). The bifurcation is one-sided — supercritical if $b > 0$ and subcritical if $b < 0$. ii) k odd. Choose either $b = +1$ or $b = -1$ and take ξ to be the positive root of (3.12). In either case the bifurcation is transcritical (the nontrivial solutions appear on both sides of criticality).

Now the solution to the full bifurcation equations are obtained as a consequence of the implicit function theorem. One fixes b and computes $g_\xi(0, b, \xi_0) = (k-1)A_{k0}\xi_0^{k-2} \neq 0$. The parametrized family $\lambda = b\epsilon^k$, $\alpha = \epsilon\xi(\epsilon)$ is analytic in ϵ. From this argument we see that the solution $\lambda, u(\alpha, \lambda)$ of the original problem can be analytically parametrized in the general case; and therefore one can always expand the non-trivial bifurcating solutions of $G(\lambda, u)$ as a power series in ϵ.

4. Stability of the Bifurcating Solutions

To determine the stability of the bifurcating solutions we study the perturbation of the critical eigenvalue σ of $G_u(\lambda, u)$ along the bifurcating branches. Assume λ, u to be parametrized so that we have to consider the operator $L(\epsilon) = G_u(\lambda(\epsilon), u(\epsilon))$. The general stability results stated in Theorem 3.1 will be derived later on (in Chapter IV) as a consequence of a general theory of bifurcation and stability at a multiple eigenvalue. For the present we compute the perturbation series for $\sigma(\epsilon)$ for the simple example

$$G(\lambda, u) = L_0 u + \lambda L_1 u + N(u, u)$$

where L_0 has a one-dimensional kernel spanned by φ_0. The Hopf condition $\sigma'(0) > 0$ $(\sigma = \sigma(\lambda))$ is then given by $< L_1\varphi_0, \varphi_0 > \; > 0$. The bifurcating solutions can be expanded in a power series

$$\lambda = \lambda_1 \epsilon + \lambda_2 \epsilon^2 + \ldots$$

$$u = \epsilon\varphi_0 + \epsilon^2 \psi_0 + \epsilon^3 \psi_1 + \ldots$$

where

$$L_0\psi_0 + \lambda_1 L_1 \varphi_0 + N(\varphi_0, \varphi_0) = 0$$

$$\lambda_1 < L_1\varphi_0, \varphi_0^* > + < N(\varphi_0, \varphi_0), \varphi_0^* > = 0.$$

Assume $\lambda_1 \neq 0$ and let $\sigma(\epsilon)$, $\varphi(\epsilon)$ be the critical eigenvalue and eigenfunction of the linearized operator. These satisfy

$$(L_0 + \lambda(\epsilon)L_1)\varphi(\epsilon) + 2N(u(\epsilon), \varphi(\epsilon)) = \sigma(\epsilon)\varphi(\epsilon).$$

Expanding out in powers of ϵ and setting the coefficient of ϵ equal to zero, we get

$$L_0\varphi_1 + \lambda_1 L_1 \varphi_0 + 2N(\varphi_0, \varphi_0) = \sigma_1 \varphi_0 .$$

Operating on this equation by φ_0^* we get

$$\sigma_1 = \lambda_1 < L_1\varphi_0, \varphi_0^* > + 2 < N(\varphi_0, \varphi_0), \varphi_0^* >$$

$$= -\lambda_1 < L_1\varphi_0, \varphi_0^* > .$$

Since $< L_1\varphi_0, \varphi_0^* > \, > 0$ for loss of stability, this tells us that

$$\frac{\sigma(\epsilon)}{\lambda(\epsilon)} = - < L_1\varphi_0, \varphi_0^* > + O(\epsilon)$$

for small ϵ and therefore that close to the bifurcation point the sign of σ is opposite to that of λ. Consequently when $\lambda_1 \neq 0$, supercritical solutions are stable and subcritical solutions are unstable.

When $\lambda_1 = 0$ but $\lambda_2 \neq 0$ we have to calculate one successive term higher in the perturbation series. For the solution $\lambda(\epsilon), u(\epsilon)$ itself, we have

$$L_0\psi_0 + N(\varphi_0, \varphi_0) = 0 \tag{3.13a}$$

$$L_0\psi_1 + \lambda_2 L_1\varphi_0 + 2N(\varphi_0, \psi_0) = 0 \tag{3.13b}$$

From the second equation we obtain

$$\lambda_2 < L_1\varphi_0, \varphi_0^* > + 2 < N(\varphi_0, \psi_0), \varphi_0^* > = 0 \tag{3.14}$$

The perturbation series for the eigenfunction and eigenvalue is

$$(L_0 + \epsilon^2\lambda_2 L_1)(\varphi_0 + \epsilon\varphi_1 + \ldots) + 2N(\epsilon\varphi_0 + \epsilon^2\psi_0 + \epsilon^3\psi_1 + \ldots, \varphi_0 + \epsilon\varphi_1 + \ldots)$$
$$= (\epsilon\sigma_1 + \epsilon^2\sigma_2 + \ldots)(\varphi_0 + \epsilon\varphi_1 + \ldots).$$

Collecting powers of ϵ, we obtain

$$L_0\varphi_1 + 2N(\varphi_0, \varphi_0) = \sigma_1\varphi_0$$

$$L_0\varphi_2 + \lambda_2 L_1\varphi_0 + 2N(\varphi_0, \varphi_1) + 2N(\psi_0, \varphi_0) = \sigma_1\varphi_1 + \sigma_2\varphi_0 \, .$$

Operating on the first equation with φ_0^* we get $\sigma_1 = 0$, since $<N(\varphi_0, \varphi_0), \varphi_0^*> = 0$ (our assumption was that $\lambda_1 = 0$). Applying φ_0^* we get

$$\lambda_2 < L_1\varphi_0, \varphi_0^* > + 2 < N(\varphi_1, \varphi_0), \varphi_0^* > + 2 < N(\varphi_0, \psi_0), \varphi_0^* > = \sigma_2$$

From this and (3.14) we see that

$$\sigma_2 = 2 < N(\varphi_1, \varphi_0), \varphi_0^* >$$

Since $\lambda_1 = \sigma_1 = 0$, φ_1 satisfies

$$L_0\varphi_1 + 2N(\varphi_0, \varphi_0) = 0$$

while ψ_0 satisfies (3.13a). Since $<\varphi_1, \varphi_0^*> = <\psi_0, \varphi_0^*> = 0$, the solution is unique and so $\varphi_1 = 2\psi_0$. Therefore

$$\sigma_2 = 4 < N(\psi_0, \varphi_0), \varphi_0^* > = -2\lambda_2 < L_1\varphi_0, \varphi_0^* > .$$

The reader can easily check that this implies the stability results of Theorem 3.1 in this simple case. A proof of the general result will be given in Chapter IV.

5. Bifurcation of Periodic Solutions

Now we turn to the so-called Hopf bifurcation theorem. Suppose $G(\lambda, u)$ vanishes identically for $u = 0$ and that this solution loses stability as λ crosses zero by virtue of a pair of complex conjugate eigenvalues crossing the imaginary axis. Under these and various technical conditions we demonstrate the existence of a bifurcating family of periodic solutions of the time dependent equation

$$\frac{\partial u}{\partial t} - G(\lambda, u) = 0 \qquad (3.15)$$

We are going to construct a one parameter family $v(s, \epsilon)$, $\omega(\epsilon)$, $\lambda(\epsilon)$ of solutions of (3.15), where v is 2π periodic in s. The function $u(t, \epsilon) = \epsilon v(\omega(\epsilon)t, \epsilon)$ is then a periodic solution of (3.15) with period $2\pi/\omega$. The method here is that of Joseph and Sattinger [4].

Let $\gamma(\lambda)$ be the critical eigenvalue of $L(\lambda) = G_u(\lambda, 0)$, so that $L(\lambda)\phi(\lambda) = \gamma(\lambda)\varphi(\lambda)$, where $\mathrm{Re}\ \gamma(0) = 0$, $\mathrm{Re}\ \gamma'(0) > 0$. For physical problems it is usually the case that the operator G is real (that is, G commutes with complex conjugation: $\overline{G(\lambda, u)} = G(\overline{\lambda}, \overline{u})$. Then taking complex conjugates, $L(\lambda)\overline{\varphi(\lambda)} = \overline{\gamma}\,\overline{\varphi}$. We may assume $\gamma(0) = i$ by a suitable choice of time units. Then, with $L_0 = L(0)$ and $\varphi_0 = \varphi(0)$,

$$L_0\varphi_0 = i\varphi_0 \quad , \quad L_0\overline{\varphi_0} = -i\overline{\varphi_0} .$$

By the Riesz-Schauder theory for the operator L_0 there exists an eigenfunction φ_0^* of the adjoint operator L_0^* such that

$$L_0^*\varphi_0^* = i\varphi_0^* \quad , \quad L_0^*\overline{\varphi_0^*} = -i\overline{\varphi_0^*} .$$

Assume i is a simple eigenvalue of L_0 with Riesz index 1. That is, there are no functions ψ such that $L_0^2\psi = 0$ while $L_0\psi \neq 0$. Then $<\varphi_0, \varphi_0^*> \neq 0$; for if $<\varphi_0, \varphi_0^*> = 0$ then $L_0 u = \varphi_0$ is solvable, which means that the solution u satisfies $L_0^2 u = 0$. Therefore we may normalize φ_0 and φ_0^* so that $<\varphi_0, \varphi_0^*> = 1$. Furthermore, $<\overline{\varphi_0}, \varphi_0^*> = 0$, for

$$-i<\overline{\varphi_0},\varphi_0^*> \;=\; <-i\overline{\varphi_0},\varphi_0^*> \;=\; <L_0\overline{\varphi_0},\varphi_0^*>$$

$$= <\overline{\varphi_0},L_0^*\varphi_0^*> \;=\; <\overline{\varphi_0},i\varphi_0^*> \;=\; i<\overline{\varphi_0},\varphi_0^*>$$

and therefore $2i<\overline{\varphi_0},\varphi_0^*> \;=\; 0.$

Let us compute the quantity $\mathrm{Re}\,\gamma'(0)$. By perturbation theory

$$\varphi(\lambda) \;=\; \varphi_0 + \lambda\varphi_1 + \dots \qquad \gamma(\lambda) = i + \lambda\gamma_1 + \dots\ .$$

Substituting these series into the eigenvalue problem we obtain

$$L_0\varphi_0 \;=\; i\varphi_0 \quad \text{and} \quad L_0\varphi_1 + L_1\varphi_0 \;=\; i\varphi_1 + \gamma_1\varphi_0\ .$$

Operating on the second equation with φ_0^* we get

$<L_1\varphi_0,\varphi_0^*> \;=\; \gamma_1<\varphi_0,\varphi_0^*> \;=\; \gamma_1$; and therefore

$$\mathrm{Re}\,\gamma_1 = \mathrm{Re}\,\gamma'(0) = \mathrm{Re}<L_1\varphi_0,\varphi_0^*> \;=\; \mathrm{Re}<G_{u\lambda}\varphi_0,\varphi_0^*>\ . \tag{3.16}$$

We now introduce the Banach spaces $\mathcal{E}_{2\pi}$ and $\mathcal{F}_{2\pi}$, where G maps $\Lambda\times\mathcal{E}$ to $\mathcal{F}$. These spaces consist of 2π periodic functions with values in $\mathcal{E}$ and $\mathcal{F}$ respectively. We assume they are endowed with an appropriate norm so that $\mathcal{E}_{2\pi}\subset\mathcal{F}_{2\pi}$ and so that the analysis descirbed below goes through. For parabolic systems (including the Navier-Stokes equations) the time dependent Hölder norms described in Chapter II are sufficient. Consider the operator

$$J_u \equiv \frac{\partial u}{\partial s} - L_0 u \;=\; \frac{\partial u}{\partial s} - G_u(0,0)u.$$

J has a two-dimensional null space $\mathcal{N}$ spanned by $e^{is}\varphi_0$ and $e^{-is}\overline{\varphi_0}$. Denote these two functions by ψ_1 and $\psi_2 = \overline{\psi_1}$. The adjoint operator

$$J^* = -\frac{\partial}{\partial s} - L_0^*$$

has null functions $\psi_1^* = e^{-is}\varphi_0^*$ and $\psi_2^* = e^{is}\varphi_0^*$.

Now consider linear functionals of the form

$$<\varphi, \psi> = \frac{1}{2\pi}\int_0^{2\pi} <\varphi(s), \psi^*(s)>_{\mathcal{E}} \, ds \tag{3.17}$$

where φ takes its values in $\mathcal{E}$ and ψ^* takes its values in $\mathcal{E}^*$. The reader may verify that in terms of this bilinear form $<\psi_i, \psi_j^*> = \delta_{ij}$ and

$$\mathrm{Re}\,\gamma'(0) = \mathrm{Re} <L_1\psi_1, \psi_1^*> . \tag{3.18}$$

Let P be the projection onto the kernel $\mathcal{N}$ of J. P takes the form

$$Pu = <u, \psi_1^*>\psi_1 + <u, \psi_2^*>\psi_2$$

where $< , >$ is the one given in (3.17). $P^2 = P$ since $<\psi_i, \psi_j^*> = \delta_{ij}$; and furthermore $PJ = JP$. Let $Q = I - P$. We make the following explicit assumption.

> Fredholm Alternative for J. Assume the equation $Ju = f$ is solvable for $f \in \mathcal{F}_{2\pi}$ whenever $<f, \psi_1^*> = <f, \psi_2^*> = 0$, and that J is an isomorphism from $Q\mathcal{E}_{2\pi}$ to $Q\mathcal{F}_{2\pi}$.

The Fredholm alternative for J is valid for a fairly general class of parabolic systems. It is verified for the Navier-Stokes equations in [4],[10].

For convenience we introduce the notation $[u] = <u, \psi_1^*>$ when $u \in \mathcal{E}_{2\pi}$. The following relationships are easily derived:

$$[\psi_1] = 1 \ , \ [\psi_2] = 0 \tag{3.19a}$$

$$[\frac{\partial u}{\partial s}] = -i[u] \tag{3.19b}$$

$$< \frac{\partial u}{\partial s}, \psi_2^* > = i < u, \psi_2^* > . \tag{3.19c}$$

When u is real-valued, $\overline{< u, \psi_1^* >} = < u, \psi_2^* >$ so $Pu = 2\mathrm{Re}\,([u]\psi_1)$.

We can choose the phase of u so that $[u] > 0$, as follows. Equations (3.15) are invariant under time translations. Let $(T_\tau u)(t) = u(t + \tau)$ and set

$$H(\lambda, \omega, u) = \omega \frac{\partial u}{\partial s} - G(\lambda, u) . \tag{3.20}$$

Then $T_\tau H(\lambda, \omega, u) = H(\lambda, \omega, T_\tau u)$. Furthermore

$$[T_\tau u] = e^{-i\tau}[u] \tag{3.21}$$

so given any solution u we can shift its phase so that $[u] > 0$.

The equation $H(\lambda, \omega, u) = 0$ can be written in the following form by expanding H in a Taylor series about $\lambda = 0$, $\omega = 1$, $u = 0$:

$$\begin{aligned} H(\lambda, \omega, u) &= H_u(0,1,0)u + H_\omega(\omega - 1) + H_\lambda \lambda + H_{\lambda u} \lambda u + R(\lambda, u) \\ &= Ju + (\omega - 1)\frac{\partial u}{\partial s} - \lambda L_1 u + R(\lambda, u) \end{aligned} \tag{3.22}$$

where $\| R(\lambda, u) = o(\| u \|)$. Note that $H_\lambda(0, 1, 0) = 0$. Now write $u = \epsilon v$ where $[v] = 1$ and expand v as

$$v = v_0 + \chi$$

where $Pv = v_0$ and $P\chi = 0$. Since $[v] = 1$, $Pv = 2\,\mathrm{Re}\,\psi_1 = \psi_1 + \psi_2$. Substituting this expression into the Taylor series for H, we obtain

$$\epsilon J\chi + \epsilon(\omega-1)\frac{\partial v_0}{\partial s} + \epsilon(\omega-1)\frac{\partial \chi}{\partial s} - \epsilon\lambda L_1 v_0 - \lambda\epsilon L_1\chi + \epsilon^2 R_1(\lambda, v_0+\chi, \epsilon) = 0$$

Since $R(\lambda, u)$ begins with quadratic terms in u, we can write $R(\lambda, \epsilon v) = \epsilon^2 R_1(\lambda, v, \epsilon)$, where R_1 is regular in λ, v, and ϵ. Dividing the above equation by ϵ we obtain

$$J\chi + (\omega-1)\frac{\partial v_0}{\partial s} + (\omega-1)\frac{\partial \chi}{\partial s} - \lambda L_1 v_0 - \lambda L_1 \chi + \epsilon R_1(\lambda, v_0+\chi, \epsilon) = 0.$$

We now apply the projections P and Q to get the system of equations

$$J\chi + (\omega-1)\frac{\partial \chi}{\partial s} - \lambda Q L_1 v_0 - \lambda Q L_1 \chi + \epsilon Q R_1(\lambda, v, \epsilon) = 0 \qquad (3.23a)$$

$$(\omega-1)<\frac{\partial v_0}{\partial s}, \psi_1^*> - \lambda<L_1 v_0, \psi_1^*> - \lambda<L_1\chi, \psi_1^*> + \epsilon<R_1, \psi_1^*> = 0 \qquad (3.23b)$$

$$(\omega-1)<\frac{\partial v_0}{\partial s}, \psi_1^*> - \lambda<L_1 v_0, \psi_2^*> - \lambda<L_1\chi, \psi_2^*> + \epsilon<R_1, \psi_2^*> = 0 \qquad (3.23c)$$

Using the relations (3.20), (3.21), we get

$$<\frac{\partial v_0}{\partial s}, \psi_1^*> = -i<v_0, \psi_1^*> = -i$$

$$<\frac{\partial v_0}{\partial s}, \psi_2^*> = i<v_0, \psi_2^*> = i \quad .$$

Furthermore

$$<L_1 v_0, \psi_1^*> = <L_1\psi_1, \psi_1^*> = \gamma'(0)$$

and

$$<L_1 v_0, \psi_2^*> = \overline{<L_1 v_0, \psi_1^*>} = \overline{\gamma'(0)} \quad .$$

We now solve the system (3.24) by applying the implicit function theorem. When $\epsilon = 0$ the system is satisfied by $\omega = 1$, $\lambda = 0$, $\psi = 0$. The Frechet derivative of the system at this point is the matrix operator

$$\begin{bmatrix} J & 0 & QL_1 v_0 \\ 0 & -i & -\gamma'(0) \\ 0 & i & \overline{-\gamma'(0)} \end{bmatrix}$$

which is an isomorphism from $Q\mathcal{E}_{2\pi} \times \mathbb{C} \times \mathbb{C}$ to $Q\mathcal{F}_{2\pi} \times \mathbb{C} \times \mathbb{C}$. By the implicit function theorem then there is an analytic solution $\lambda = \lambda(\epsilon)$, $\omega = \omega(\epsilon)$, $\chi = \chi(\epsilon)$, and this family constitutes a bifurcating time periodic solution of (3.15).

Given that such a power series solution exists we can make the substitution into $H(\lambda, \omega, u) = 0$ and compute the successive terms in the expansion. We have

$$\lambda = \epsilon\lambda_1 + \epsilon^2\lambda_2 + \ldots$$

$$\omega = 1 + \epsilon\omega_1 + \epsilon^2\omega_2 + \ldots$$

$$u = \epsilon v_0 + \epsilon^2(\chi_0 + \chi_1 + \ldots)$$

Let us carryout the computations for the case of a quadratic nonlinearity $G(\lambda, u) = (L_0 + \lambda_1 L_1)u + N(u, u)$.

Substituting these expressions into (3.24) and collecting powers of ϵ, we get at order ϵ^2

$$J\chi_0 + \omega_1 \frac{\partial v_0}{\partial s} - \lambda_1 L_1 v_0 - N(v_0, v_0) = 0 \qquad (3.24)$$

In order to solve this for ψ_0 we must choose ω_1 and λ_1 so that

$$<\omega_1 \frac{\partial v_0}{\partial s} \ , -\lambda_1 L_1 v_0 - N(v_0, v_0), \psi_j^* > \ = \ 0 \qquad j = 1, 2.$$

It is easily seen that the term $<N(v_0, v_0), \psi_j^*> \ = \ 0$ since this expression reduces to a trigonometric polynomial of odd order. Therefore the above condition reduces to

$$i\omega_1 + \gamma'(0)\lambda_1 \ = \ 0 \quad , \quad -i\omega_1 + \overline{\gamma'(0)}\lambda_1 \ = \ 0 \ .$$

These equations are uniquely solvable, giving $\lambda_1 = \omega_1 = 0$. At the next stage we get

$$J\chi_1 + \omega_2 \frac{\partial v_0}{\partial s} - \lambda_2 L_1 v_0 - 2N(v_0, \psi_0) = \ 0 \ .$$

Operating by ψ_1^* and ψ_2^* this time we get

$$i\,\omega_2 + \gamma'(0)\lambda_2 \ + \ 2<N(v_0, \chi_0), \ \psi_1^*> \ = \ 0$$

$$-i\omega_2 + \gamma'(0)\lambda_2 + \ 2<N(v_0, \chi_0), \ \psi_2^* > \ = \ 0$$

These two equations are easily solved for ω_2 and λ_2. We get

$$\lambda_2 \mathrm{Re}\, \gamma'(0) \ + \ 2<N(v_0, \chi_0), \mathrm{Re}\ \psi_1^* > = 0 \tag{3.25a}$$

$$\omega_2 + \lambda_2 \mathrm{Im}\ \gamma'(0) + 2<N(v_0, \chi_0), \mathrm{Im}\, \psi_1^* > \ = 0 \tag{3.25b}$$

6. Stability: Floquet exponents

In the case of ordinary differential equations the stability of a periodic solution of an autonomous system

$$\dot{x} = F(x)$$

can be determined by computing the Floquet exponents of the variation equations

$$\dot{y} = F'(x(t))y \quad \text{or} \quad \dot{y}_i = \frac{\partial F_i}{\partial x_j}(x(t))y_j \ ,$$

where $x(t)$ is the periodic solution in question, and $F'(x(t))$ denotes the Frechet derivative (Jacobian) of F at $x(t)$.

The Floquet exponents are determined as follows: Given a linear system of ordinary differential equations with periodic coefficients

$$y = A(t)y \ , \quad \text{where} \quad A(t) = A(t+T) \tag{3.26}$$

denote by $\Phi(t)$ the fundamental solution matrix:

$$\dot{\Phi}(t) = A(t)\Phi \quad \text{and} \quad \Phi(0) = I \ .$$

The eigenvalues of $\Phi(T)$ are called the Floquet multipliers of (3.26). If μ is a multiplier, then there is a vector ψ such that $\Phi(T)\psi = \mu\psi$. Equivalently, if $\dot{z}(t) = A(t)z$ and $z(0) = \psi$, then $z(T) = \mu\psi = \mu z(0)$. Putting $\mu = e^{\beta T}$ and $w(t) = e^{-\beta t}z(t)$, then

$w(T) = e^{-\beta T}z(T) = \mu^{-1}z(T) = z(0) = w(0)$,

$$\dot{w} + \beta w - A(t)w = 0 .$$

Thus the Floquet exponent β can be considered an eigenvalue of the problem

$$w - A(t) + \beta w = 0$$
$$w(0) = w(T) \tag{3.27}$$

If $\operatorname{Re}\beta < 0$ then $|\mu| < 1$ and $|z(T)| < |z(0)|$. If all the Floquet exponents have negative real parts, then all the eigenvalues of $\Phi(T)$ are less than one in absolute value, and $\Phi(T)$ is a contraction mapping. In the case of a periodic solution $z(t)$ of an autonomous system, however, $y(t) = \dot{x}(t)$ is a periodic solution of the variation equations; hence one of the Floquet multipliers is always unity. Nevertheless, if all the remaining Floquet multipliers of the variational equation are less than one in absolute value, the periodic solution $x(t)$ is orbitally stable. (See Coddington and Levinson [1])

To be precise, let $y(t, y_0)$ denote the solution of $\dot{y} = F(y)$, $y(0) = y_0$. Then $x(t)$ is orbitally stable if there is a number $\epsilon > 0$ such that whenever $|y_0 - x(0)| < \epsilon$ then $|y(t, y_0) - x(t+\delta)| \to 0$ as $t \to \infty$ for some number δ $(0 \le \delta < T)$.

The eigenvalue problem for the Floquet exponents corresponding to (3.15) is

$$\frac{\partial w}{\partial t} - G_u(\lambda, u)w + \beta w = 0 \qquad (3.28)$$

$$w(0) = w\left(\frac{2\pi}{\omega}\right).$$

Recall that $\lambda = \tau(\epsilon)$, $\omega(\epsilon)$ and $v(\epsilon, s)$ are predetermined. We change variables by putting $s = \omega t$ and replace β by $\omega\beta$; then we get

$$\omega \frac{\partial w}{\partial s} - G_u(\lambda, u)w + \beta\omega w = 0 \qquad (3.29)$$

$$w(0) = w(2\pi).$$

One solution of (3.29) is $\beta = 0$ and $w = \frac{\partial u}{\partial s}$; this solution is obtained by differentiating (3.22) with respect to s. When $\epsilon = 0$, $\omega = 1$ and $\lambda = 0$ we have the eigenvalue problem

$$\frac{\partial w}{\partial s} - L_0 w + \beta w = 0$$

$$w(0) = w(2\pi)$$

Solutions of this problem are of the form $w(x, s) = e^{-\beta s}\varphi(x)$, where $\beta = \mu + 2\pi in$, φ is an eigenfunction of L_0 with eigenvalue μ, and n is an arbitrary integer. Under the bifurcation assumption that L_0

has eigenvalues at $\pm i$ and all other eigenvalues in the left half plane, we see that when $\epsilon = 0$ (3.29) has double eigenvalues at $\pm 2\pi$ in and all remaining eigenvalues in the left half plane.

As we have seen, one Floquet exponent of (3.29) is always $\beta = 0$ with $w = \frac{\partial u}{\partial s}$ (for small $\epsilon \neq 0$). We construct the other Floquet function in the form

$$w = a(\epsilon)\frac{\partial v}{\partial s} + \varphi , \tag{3.30}$$

where $P\varphi = 0$. Let us treat the case of a quadratic nonlinearity, $G(\lambda, u) = (L_0 + \lambda L_1)u + N(u, u)$. Recall that $u = \epsilon v$ and $v = v_0 + \chi$. The function $\frac{\partial v}{\partial s}$ satisfies

$$\omega \frac{\partial^2 v}{\partial s^2} - L(\lambda)\frac{\partial v}{\partial s} - 2\, N(v, \frac{\partial v}{\partial s}) = 0 \tag{3.31}$$

Substituting the expression (3.30) for w into (3.29) and using (3.31) we get

$$\begin{aligned} &\omega \frac{\partial \varphi}{\partial s} - L(\lambda)\varphi - 2\epsilon N(v, \varphi) + \beta\omega\varphi + \beta\omega a \frac{\partial v}{\partial s} = 0 \\ &\psi(0) = \psi(2\pi) \end{aligned} \tag{3.32}$$

We seek $\psi(s, \epsilon)$, $\beta(\epsilon)$ and $a(\epsilon)$ in the form

$$\varphi = u_0 + \epsilon\eta \ , \quad [\eta] = 0$$

$$\beta(\ ) = \epsilon^2\sigma(\epsilon) \ , \quad \sigma = \sigma_0 + \epsilon\sigma_1 + \dots$$

$$a(\epsilon) = a_0 + \epsilon a_1 + \dots \ .$$

The convergence of these power series can be established by using the implicit function theorem as we did in the previous section to prove the existence of the periodic motions themselves. The details are given in Sattinger [10].

Assuming the convergence of the power series, then, it remains to calculate the successive terms, and in particular the relationship between σ_0 and λ_2, since this will show that the bifurcating periodic solutions are stable or unstable according as the bifurcation is supercritical or subcritical. We continue to consider the case of a quadratic nonlinearity. Substituting (3.30) into (3.29) and using (3.31) we obtain

$$\omega\frac{\partial\varphi}{\partial s} - L(\lambda)\varphi - 2\ N(v,\varphi) + \beta\omega\varphi \;=\; -\beta\omega a\,\frac{\partial v}{\partial s}\,.$$

Expanding $\varphi, \omega, \lambda, \beta, a$ in their Taylor series in ϵ and equating the coefficient of each power of ϵ to zero, we get successively

$$Jv_0 = 0 \tag{3.33a}$$

$$J\eta_0 - 2N(v_0, v_0) = 0 \tag{3.33b}$$

$$J\eta_1 + \omega_2\frac{\partial v_0}{\partial s} - \lambda_2 L_1 v_0 - 2N(\chi_0, v_0) - 2N(v_0, \eta_0) + \sigma_0 v_0 = -\sigma_0 a_0\,\frac{\partial v_0}{\partial s}\,. \tag{3.33c}$$

The first equation is automatically satisfied; and the second is solvable for η_0, since, as we have already noted, $\langle N(v_0, v_0), \psi_j^*\rangle = 0$ for $j = 1, 2$. The solution of (3.33b) is uniquely determined if we require $[\eta_0] = 0$. Comparing (3.33b) with (3.24) and recalling that $\omega_1 = \lambda_1 = 0$ we see that $\eta_0 = 2\chi_0$. Going now to (3.33c) and operating on that equation with ψ_1 we get

$$-i\omega_2 - \lambda_2[L_1 v_0] - 6[N(\chi_0, v_0)] + \sigma_0 \;=\; i\,\sigma_0 a_0\,.$$

Now $[L_1 v_0] = \gamma'(0)$ and the normalization of the Floquet function w can be chosen so that $a_0 = 1$. Taking imaginary parts of this equation we get

$$\omega_2 + \lambda_2 \operatorname{Im}\gamma'(0) + 6 \langle N(\chi_0, v_0), \operatorname{Im} \psi_1^* \rangle = \sigma_0 .$$

If we compare the above equation with (3.25b) we see that

$$\sigma_0 = 4 \langle N(\chi_0, v_0), \operatorname{Re} \psi_1^* \rangle ;$$

and from (3.25a) we get

$$\sigma_0 = -2(\operatorname{Re}\gamma'(0))\lambda_2 .$$

This relationship tells us that supercritical solutions are orbitally stable because $\beta < 0$ and subcritical solutions are unstable (at least according to the principle of linearized stability); for if $\lambda_2 > 0$ the bifurcation is supercritical and $\sigma_0 < 0$, indicating that the critical Floquet exponent moves to the left when the periodic solutions branch to the right, and vice versa.

Remarks: The device (3.30) was suggested by D. D. Joseph and was used in the original paper on the bifurcation of periodic motions of the Navier-Stokes equations by Joseph and myself [4]. The rigorous relationship between the Floquet exponents and the orbital stability has been established by Iudovich and Iooss (see [2] and [6]).

7. Uniqueness of the Periodic Motions

In this section we show that locally, in the neighborhood of the bifurcation point, the bifurcating periodic motions are unique up to a phase shift. We make the following compactness assumption:

If $\{\lambda_n, \omega_n, u_n\}$ is a bounded sequence of solutions of $H(\lambda, \omega, u) = 0$, it contains a convergent subsequence. This is certainly the case for the Navier-Stokes equations or for more general classes of parabolic systems. For one may expect that the time periodic motions obtained are smooth and bounded in any Holder norm $\| \quad \|_{2, 2\alpha; 1, \alpha}$, say, for $0 < \alpha \leq \frac{1}{2}$; one may then choose for the space $\mathcal{E}_{2\pi}$ a Hölder space of any $\alpha < \frac{1}{2}$, and use the fact that C_α is compactly embedded in $C_{\alpha'}$ when $\alpha < \alpha'$. Similarly, the compactness assumption holds for systems of ordinary differential equations with sufficiently smooth (say C^k) nonlinear terms; for then the solutions obtained are a priori bounded not only in the C^1 norm, but in higher order norms as well.

Now suppose $(\lambda_n, \omega_n, u_n)$ is a sequence of periodic solutions of $H(\lambda, \omega, u) = 0$ which is uniformly bounded and tends, as $n \to \infty$, to $(0, \omega_0, 0)$, where ω_0 is not necessarily equal to one. We show that for sufficiently large n this sequence lies on some phase shift of the bifurcating family of periodic motions which we have already constructed. First, by a suitable phase shift, we adjust u_n so that $[u_n] > 0$. Putting

$\epsilon_n = \|u_n\|$ and $u_n = \epsilon_n v_n$ we have $\|v_n\| = 1$ and $[v_n] > 0$. Substituting $u_n = \epsilon_n v_n$ into (3.22) and dividing out by ϵ_n, we get

$$Jv_n + (\omega_n - 1)\frac{\partial v_n}{\partial s} - \lambda_n L_1 v_n + \epsilon_n R_1(\lambda_n, v_n, \epsilon_n) = 0. \tag{3.34}$$

Now applying our compactness hypothesis we choose a convergent subsequence $v_{n'}$, which we also denote by v_n, so that $v_n \to f$ as $n \to \infty$. Taking limits as $n \to \infty$ in (3.34) we get

$$Jf + (\omega_0 - 1)\frac{\partial f}{\partial s} = 0, \quad \text{or} \quad \omega_0 \frac{\partial f}{\partial s} - L_0 f = 0.$$

Now $f \neq 0$ since $\|f\| = \lim_{n\to\infty} \|v_n\|$; and the only nontrivial solution of the above equation is $\omega_0 = 1$ and $f = a\psi_1 + b\psi_2$. Furthermore, since $[v_n] > 0$, it follows that $[f] > 0$ and therefore that $a = b$ and f is a scalar multiple of v_0. Thus if we write $v_n = \alpha_n v_0 + \chi_n$, we see that $\chi_n \to 0$ and $\alpha_n \to \|v_0\|^{-1}$ as $n \to \infty$. Since $\alpha_n \not\to 0$ we may divide by α_n and change our scaling; we replace v_n by v_n/α_n and χ_n by χ_n/α_n. Then writing $u_n = \epsilon_n v_n$, we have $\epsilon_n = [u_n]$, since $[v_n] = 1$; and v_n now has the form $v_n = v_0 + \chi_n$ where χ_n still tends to zero as $n \to \infty$. Substituting this form of u_n into (3.22) we obtain

$$J\chi_n + (\omega_n - 1)\frac{\partial v_0}{\partial s} + (\omega_n - 1)\frac{\partial \chi_n}{\partial s}$$
$$- \lambda_n L_1 v_0 - \lambda_n L_1 \chi_n + \epsilon_n R_1(\lambda_n, v_0 + \chi_n, \epsilon_n) = 0.$$

Operating on this equation respectively by the projection Q and the linear functionals ψ_1 and ψ_2 we obtain for $(\lambda_n, \epsilon_n, \omega_n, \chi_n)$ the same set of equations (3.23) which we previously solved to obtain the bifurcating family of periodic motions. Since the solutions to these equations was obtained as a consequence of the implicit function theorem, it is unique in a neighborhood of $(0, 0, 1, 0)$. Therefore, for large n, $(\lambda_n, \epsilon_n, \omega_n, \chi_n)$ coincides with the solution already obtained, and the uniqueness of the bifurcating periodic motions is established.

We close this chapter with a statement of the general theorem.

Theorem 3.2. Let $G(\lambda, u)$ be an analytic mapping from $\Lambda \times \mathcal{E}$ to $\mathcal{F}$ where $\mathcal{E} \subset \mathcal{F}$ are complex Banach spaces. Let $G(\lambda, 0) \equiv 0$ and suppose the zero solution loses stability as λ crosses zero by virtue of a pair of simple complex conjugate eigenvalues $\gamma(\lambda)$ and $\overline{\gamma}(\lambda)$ crossing the imaginary axis: $\gamma(0) = i\omega_0$, $\operatorname{Re} \gamma'(0) > 0$. Let the operator

$$J = \omega_0 \frac{\partial}{\partial s} - L_0$$

be a Fredholm operator mapping the Banach space $\mathcal{E}_{2\pi}$ to $\mathcal{F}_{2\pi}$, these being Banach spaces of 2π periodic functions with values in $\mathcal{E}$ and $\mathcal{F}$ respectively. Then there exists an analytic one parameter family of periodic solutions of (3.1) of the form $\lambda(\epsilon), \omega(\epsilon), v(\omega(\epsilon)t, \epsilon)$ bifurcating from the origin, with $\lambda(0) = 0$ and $\omega(0) = \omega_0$.

These solutions are stable when they branch supercritically and unstable when they branch subcritically. They are unique up to phase shifts if one makes the compactness assumption that any bounded sequence of periodic solutions of (3.1) (with bounded periods) contains a convergent subsequence.

References

1. E. A. Coddington and N. Levinson, Theory of Ordinary Differential Equations, McGraw-Hill, New York, 1955.

2. G. Iooss, "Existence et stabilité de la solution périodique secondaire intervenant dans les problèmes d'évolution du type Navier-Stokes", Arch. Rat. Mech. Anal. 47, (1972), 301-329.

3. __________, "Sur la deuxième bifurcation d'une solution stationnaire de systèmes du type Navier-Stokes", Archive Rat. Mech. Anal. 64 (1977), 339-369.

4. D. Joseph and D. H. Sattinger, "Bifurcating Time Periodic Solutions and Their Stability", Arch. Rat. Mech. Anal. 45 (1972), 75-109.

5. D. Joseph, Nonlinear Stability of Fluid Motions, Springer, New York 1976.

6. V. Judovic, "On the stability of self-oscillations of a fluid", Sov. Math. Dokl. 11 (1970), 1473-1477.

7. __________, "Appearance of auto-oscillations in a fluid", Prikl. Math. Mek. 35 (1971), 638-655.

8. D. H. Sattinger, "Bifurcation of periodic solutions of the Navier-Stokes equations", Arch. Rat. Mech. Anal. 41 (1971), 68-80.

9. __________, "Stability of bifurcating solutions by Leray-Schauder degree", Arch. Rat. Mech. Anal. 43 (1971), 154-166.

10. __________, Topics in Stability and Bifurcation Theory, Springer Lecture Notes in Mathematics, 309, 1973.

Chapter IV

BIFURCATION AT MULTIPLE EIGENVALUES

1. Lyapounov-Schmidt Procedure

We begin by discussing the general method for reducing the infinite dimensional problem to a finite dimensional one. Consider the equation (1.1) where, as usual, G is an analytic mapping from $\Lambda \times \mathcal{E}$ to $\mathcal{F}$; and suppose $G(0,0) = 0$, and that $L_0 = G_u(0,0)$ is a Fredholm operator of index zero with kernel $\mathcal{N}$ of dimension n. We assume $\mathcal{E} \subset \mathcal{F}$ so that the projection P onto the kernel $\mathcal{N}$ maps $\mathcal{E}$ into $\mathcal{E}$ and $\mathcal{F}$ into $\mathcal{F}$. Let $Q = I - P$ be the projection onto the range of L_0 in $\mathcal{F}$. The projection P was constructed in Chapter II.

We decompose (1.1) into the system

$$QG(\lambda, v+\psi) = 0 \tag{4.1}$$

$$PG(\lambda, v+\psi) = 0 \tag{4.2}$$

where $v = Pu$ and $\psi = Qu$. The first equation is solved for $\psi = \psi(\lambda, v)$ by applying the implicit function theorem. Letting $H(\lambda, v, \psi) = QG(\lambda, v+\psi)$ we note that $H(0,0,0) = 0$ and $H_\psi(0,0,0) = QL_0$. Since QL_0 is an isomorphism from $Q\mathcal{E}$ to $Q\mathcal{F}$ we can apply the implicit function theorem and conclude that (4.1) has locally a unique solution $\psi(\lambda, v)$ which is analytic in λ and v. We then substitute this solution in the second equation (4.2) to obtain the bifurcation equations

$$F(\lambda, v) \equiv PG(\lambda, v + \psi(\lambda, v)) = 0 \tag{4.3}$$

Solutions of equation (4.3) are locally in one-to-one correspondence with solutions of the system (4.1, 4.2), hence with solutions of the original equation (1.1) in the neighborhood of the bifurcation point (0, 0). The reduction procedure we have just described is known as the Lyapounov-Schmidt procedure. The solution of the infinite dimensional problem is now reduced to an algebraic problem of finding all solutions of (4.3). In addition, we shall want to determine the stability of the bifurcating solutions. The rest of this chapter will be devoted to methods of attacking equations (4.3).

2. Newton Diagrams; Reduced Bifurcation Equations

Throughout these notes we always denote the bifurcation equations by $F(\lambda, v)$. Since the original mapping G is analytic in λ and u, the bifurcation mapping F is analytic in λ and v. Here v is a vector inthe finite dimensional vector space $\mathcal{N}$. Let us expand F in a power series of multilinear operators in v:

$$F(\lambda, v) = A(\lambda)v + B_2(\lambda; v, v) + B_3(\lambda; v, v, v) + \dots \tag{4.4}$$

where $A(\lambda)$ is a linear transformation from $\mathcal{N}$ to $Q\mathcal{F}$ and

$B_k(\lambda; v, \ldots)$ is a k-linear operator from $\mathcal{N} \times \ldots \times \mathcal{N}$ to $Q\mathcal{F}$. Note that $v = 0$ is always a solution. This corresponds to our assumption that zero is always a solution of (1.1).

We now introduce a device known as the Newton diagram . It was first applied to bifurcation problems by L. Graves [2]. We rewrite the expansion of (4.4) as a power series in λ and v

$$F(\lambda, v) = \sum_{k=1}^{\infty} \sum_{i+j=k} F_{ij}(v)\lambda^j \tag{4.5}$$

where $F_{ij}(v)$ is an operator which is homogeneous of degree i: $F_{ij}(\tau v) = \tau^i F_{ij}(v)$. We now plot every point (i, j) on the lattice of nonnegative integer points in the first quadrant of the x-y plane for which F_{ij} does not vanish, as in Figure 4.1 below.

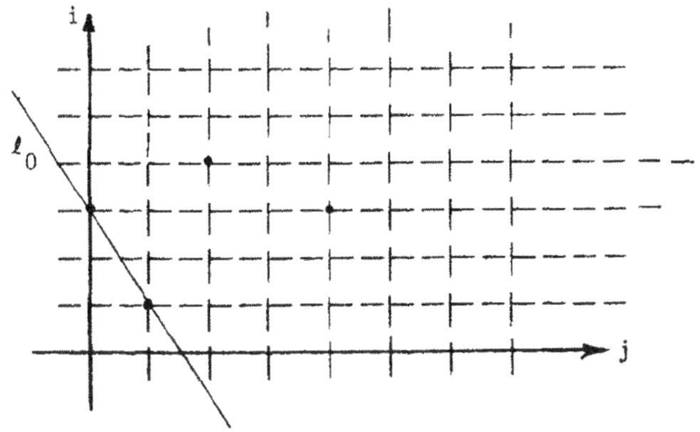

Figure 4.1

Now using the Newton diagram, we introduce a new scaling of the bifurcation equations. Let ℓ_0 be the line through points of the diagram which lies lowest to the left, and suppose its equation to be

$$ai + bj = c .$$

All other points of the diagram lie above and to the right. Let ℓ_k be the line parallel to ℓ_0 with equation

$$ai + bj = c + k.$$

We may rewrite F in the form

$$F(\lambda, v) = \sum_{k=0}^{\infty} \sum_{ai+bj=c+k} F_{ij}(v)\lambda^j .$$

We now scale the variables by introducing a new parameter ϵ and setting

$$v = \epsilon^a \xi \qquad \lambda = \epsilon^b \tau .$$

Substituting these into F we obtain

$$F(\epsilon^b \tau, \epsilon^a \xi) = \sum_{k=0}^{\infty} \epsilon^{c+k} \sum_{ai+bj=c+k} F_{ij}(\xi)\tau^j$$

$$= \epsilon^c Q(\xi, \tau) + \epsilon^{c+1} g(\xi, \tau, \epsilon).$$

Dividing by ϵ^c and letting $\epsilon \to 0$ we arrive at the reduced bifurcation equations

$$Q(\xi, \tau) = 0.$$

One would like to obtain as much information as possible from an analysis of the reduced bifurcation equations. If (ξ_0, τ_0) is a solution of the reduced bifurcation equations and $Q_\xi(\xi_0, \tau_0)$ is invertible, then a solution to the full equations can be obtained by an application of the implicit function theorem. (Keep τ_0 fixed.) The use of the Newton diagram facilitates the determination of a correct scaling for the variables v and λ. For further discussion of Newton diagrams and their application to bifurcation theory see Sather [4] and Vainberg and Trenogin [7].

We now make a number of special assumptions about the bifurcation problem. In particular, we assume $G(\lambda, 0) \equiv 0$ and that $G(0, u)$ does not vanish identically on some punctured neighborhood of the origin. We also assume a multi-dimensional analogue of the Hopf condition that the critical eigenvalue crosses strictly. First we prove

Lemma 4.1. The function $\psi(\lambda, v)$ which is determined by (4.2) satisfies the following conditions:

$$\psi(0,0) = 0\ , \quad \psi_\lambda(0,0) = 0\ , \quad \psi_v(0,0) = 0 \tag{4.6}$$

Proof. When $\lambda, v = 0$ in (4.2), we have $QG(0, \psi) = 0$; since $QG(0, 0) = 0$, $\psi = 0$ is the unique solution of this equation, so that $\psi(0, 0) = 0$. To get the second relation we simply note that since $G(\lambda, 0) \equiv 0$, $\psi(\lambda, 0)$ also vanishes identically. In fact, $QG(\lambda, \psi(\lambda, 0)) = 0$ and $QG(\lambda, 0) = 0$, so the uniqueness of the solution implies $\psi(\lambda, 0) = 0$, hence that $\psi_\lambda(0, 0) = 0$.

Finally, to get the third relation in (4.5) differentiate (4.2) with respect to v and set $\lambda, v = 0$. We get

$$QG_u(0,0)(v + \psi_v(0,0)v) = 0$$

for all $v \in \mathcal{N}$. Now $QG_u(0,0)v = 0$ and so $QG_u(0,0)\psi_v(0,0)v = 0$ for all $v \in \mathcal{N}$. This tells us that $\psi_v(0,0)v \in \mathcal{N}$ for any $v \in \mathcal{N}$; but since ψ maps $\mathcal{N}$ into $Q\mathcal{E}$ we must have $\psi_v(0,0)v = 0$, and hence $\psi_v(0,0) = 0$.

Now let us analyze the terms of lowest order in (4.5) The lowest order term is $A(0)v = F_{10}(v)$. From (4.3) we have

$$\begin{aligned} F(0,v) &= PG(0, v + \psi(0,v)) \\ &= PG_u(0,0)[v + \psi_v(0,0)v] + O(\|v\|^2) \\ &= O(\|v\|^2) ; \end{aligned}$$

hence $A(0)v = F_{10}(v) = 0$.

The next term to be considered is $F_{11}(v)\lambda$; this corresponds to the term $F_{\lambda v}(0,0)\lambda v$ in the bifurcation equations. Differentiating first with respect to v we get

$$\begin{aligned} F_v(\lambda,0)v &= PG_u(\lambda, \psi(\lambda,0))(v + \psi_v(\lambda,0)v) \\ &= PG_u(\lambda,0)(v + \psi_v(\lambda,0)v) . \end{aligned}$$

Now differentiating with respect to λ we get

$$\begin{aligned} F_{v\lambda}(0,0)\lambda v &= PG_{u\lambda}(0,0)(v + \psi_v(0,0)v)\lambda \\ &\quad + PG_u(0,0)\psi_{v\lambda}(0,0)v . \end{aligned}$$

But $PG_u(0,0) = 0$ and $\psi_v(0,0) = 0$, so

$$F_{\lambda v}(0,0)\lambda v = PG_{u\lambda}(0,0)\lambda v ,$$

or

$$F_{\lambda v}(0,0) = PG_{u\lambda}(0,0)P .$$

If we expand $G_u(\lambda, 0) = L_0 + \lambda L_1 + \ldots$ then $G_{u\lambda}(0,0) = L_1$ and

$$F_{\lambda v}(0,0) = PL_1P .$$

The matrix of $F_{\lambda v}(0,0)$ with respect to the bases $\{\varphi_1, \ldots, \varphi_n\}$ and $\{\varphi_1^*, \ldots, \varphi_n^*\}$ is

$$<F_{\lambda v}(0,0)\varphi_i, \varphi_j^*> = <L_1\varphi_i, \varphi_j^*>$$

since $<PL_1P\varphi_i, \varphi_j^*> = <L_1\varphi_i, \varphi_j^*>$. This term corresponds to the lattice point (1,1) on the Newton diagram. We shall see later that this same term is associated with the perturbation of the multiple critical point at the origin. Recall in the case of bifurcation at a simple eigenvalue we assumed the Hopf condition

$$\sigma(0) \neq 0.$$

This condition ensured that the critical eigenvalue crossed through the origin with non-zero speed. The corresponding condition for a multiple eigenvalue is that the finite dimensional linear operator PL_1P be non-singular. This will come out below when we develop the perturbation theory of the critical eigenvalues.

The last term to be analyzed in (4.5) is the first non-vanishing term $F_{m0}(v)$. Since we have assumed that $G(0,u)$ does not vanish in some punctured neighborhood of the origin we know that $F(0,v)$ does not vanish identically. Now

$$F(0,v) = \sum_{k=1}^{\infty} F_{k0}(v)$$

so there must be a first non-vanishing term $F_{m0}(v)$. Plotting i vertically and j horizontally we see that there is a point on the Newton diagram at $j = 0$, $i = m$. We shall also, in these notes, assume that the operator PL_1P is non-singular.

Under the present assumptions the line ℓ_0 has the equation

$$i + (m-1)j = m$$

The correct scaling is therefore

$$v = \epsilon\xi \qquad \lambda = \epsilon^{m-1}\tau ,$$

and the reduced bifurcation equations take the form

$$Q(\tau;\xi) = \tau A\xi + F_{m0}(\xi) = 0$$

where F_{m0} is a homogeneous operator of degree m. It suffices to restrict τ to the values $\tau = \pm 1$ since any other values can be scaled into the parameter ϵ. If m is even then $\lambda = \epsilon^{m-1}\tau$ takes on both positive and negative values. In this case all the bifurcating curves are transcritical — that is, they appear on both sides of criticality. If m is odd then $\lambda = \epsilon^{m-1}\tau$ takes on only positive or negative values, depending on the sign of τ. Thus if m is odd and (τ_0, ξ_0) is a solution of (4.7) with $\tau_0 > 0$, the bifurcation is supercritical; whereas the bifurcation is subcritical if $\tau_0 < 0$.

Now let (τ_0, ξ_0) be a solution of (4.7) and suppose we want to continue this solution to a solution of the full bifurcation equations. We examine the Frechet derivative $Q_\xi(\tau_0, \xi_0)$. If this operator is invertible then by the implicit function theorem, there is an analytic solution $(\tau_0, \xi_0(\epsilon))$ such that $\xi(0) = \xi_0$. The solution of the original problem is then $\tau_0 \epsilon^{m-1}$, $\epsilon \xi(\epsilon)$. We cannot expect in general that Q_ξ is going to be invertible; and in fact, when a continuous transformation group is present as a symmetry group of the equations, Q_ξ is definitely not going to be invertible. We shall discuss this situation later on.

3. Stability of Bifurcating Solutions

In this section we show that the stability of the bifurcating solutions is determined, to lowest order, by the eigenvalues of the Jacobian $Q_\xi(\tau_0, \xi_0)$ of the reduced bifurcation equations. We first establish the following preliminary result in the perturbation theory of an isolated n-fold eigenvalue.

Lemma 4.2. Let $L(\epsilon) = L_0 + \epsilon L_1 + \dots$ be an analytic family of bounded operators from $\mathcal{E}$ to $\mathcal{F}$ with $\mathcal{E} \subseteq \mathcal{F}$ and suppose that L_0 has an n-fold semi-simple eigenvalue at the origin. Then there is an analytic projection valued operator $E(\epsilon)$ which commutes with $L(\epsilon)$ and whose range, therefore, is invariant under $L(\epsilon)$. Let the kernel $\mathcal{N}$ of L_0 be spanned by $\{\varphi_1, \dots, \varphi_n\}$. The subspace spanned by $\{E(\epsilon)\varphi_1, \dots, E(\epsilon)\varphi_n\}$ is therefore invariant under $L(\epsilon)$, and there is an analytic matrix $B_{ij}(\epsilon)$ such that

$$L(\epsilon)E(\epsilon)\varphi_i = \sum_{j=1}^{n} B_{ij}(\epsilon)E(\epsilon)\varphi_j \tag{4.9}$$

Let $P = E(0)$ be the projection onto $\mathcal{N}$. If $PL(\epsilon) = O(\epsilon^k)$ then $L(\epsilon)E(\epsilon) = O(\epsilon^k)$ also, and the matrix $B_{ij}(\epsilon)$ coincides up to order ϵ^k with that of $PL(\epsilon)P$. Specifically, if $PL_1 = \dots = PL_{k-1} = 0$ then $PE_1 = \dots = PE_{k-1} = 0$ and

$$B_{ij}(\epsilon) = \epsilon^k < L_k\varphi_i, \varphi_j^* > + O(\epsilon^{k+1}) \tag{4 10}$$

where the vectors $\varphi_j \epsilon \mathcal{F}^*$ form a dual basis of adjoint null functions.

Proof. The projections $E(\epsilon)$ are constructed from the residue formula

$$E(\epsilon) = \frac{1}{2\pi i}\int_C (z - L(\epsilon))^{-1} dz$$

which is valid for sufficiently small ϵ; the contour C encloses the origin but no other points in the spectrum of L_0. For small ϵ the eigenvalues vary continuously so the above expression is well-defined. (See Dunford and Schwartz [1], vol. I, lemma 6, p. 586; $E(\epsilon)$ is a projection by Theorem 10, p. 568.) From the residue formula, it is immediate that $E(\epsilon)$ commutes with $L(\epsilon)$; therefore $\{E(\epsilon)\varphi_1, \ldots, E(\epsilon)\varphi_n\}$ is invariant under $L(\epsilon)$, and the existence of the matrix $B_{ij}(\epsilon)$ and the validity of equation (4.9) follow. Moreover, $B_{ij}(\epsilon)$ may be determined by solving

$$<L(\epsilon)E(\epsilon)\varphi_i, \varphi_k^*> = \sum_{j=1}^{n} B_{ij}(\epsilon)<E(\epsilon)\varphi_j, \varphi_k^*> . \tag{4.11}$$

These equations are invertible for small ϵ since

$$\det \| <E(\epsilon)\varphi_j, \varphi_k^*> \| = 1 + O(\epsilon).$$

The spectrum of $L(\epsilon)$ restricted to the invariant subspace $E(\epsilon)\mathcal{H}$ consists precisely of the eigenvalues of $B_{ij}(\epsilon)$.

Now let us show that $PL_0 = \ldots = PL_{k-1} = 0$ implies $PE_1 = \ldots = PE_{k-1} = 0$. We use the notation $E^{[j]}(\epsilon)$ to denote the jth derivative of E with respect to ϵ, and $E_j = E^{[j]}(0)$. Letting $R_z(\epsilon) = (z - L(\epsilon))^{-1}$ we have

$$E(\epsilon) = \frac{1}{2\pi i}\int_C R_z(\epsilon)dz$$

$$E^{[1]}(\epsilon) = \frac{1}{2\pi i}\int_C R_z^{\,2}(\epsilon)L^{[1]}(\epsilon)dz$$

$$E^{[k-1]}(\epsilon) = \frac{1}{2\pi i}\int_C R_z^{\,2}(\epsilon)L^{[k-1]}(\epsilon) + \ldots + (k-1)!\, R_z^{\,k}(\epsilon)(L^{[1]})^{(k-1)}dz.$$

From the above differentiation formulae it follows that

$$PE_\ell = PE^{[\ell]}(0) = \frac{1}{2\pi i}\int_C R_z^{\,2}(0)PL_\ell + \ldots + (\ell-1)!\, R_z^{\,\ell}(0)P(L_1)^\ell\, dz$$

$$= 0 \quad \text{for } \ell = 1, 2, \ldots, k-1.$$

Applying the projection P to (4.9) and using what we have just shown, we get

$$\epsilon^k(PL_kP\varphi_i) + O(\epsilon^{k+1}) = \sum_{j=1}^n B_{ij}(\epsilon)(P + \epsilon^k PE_k)\varphi_j$$

$$= \sum_{j=1}^n((B_0)_{ij} + \epsilon(B_1)_{ij} + \ldots + \epsilon^k(B_k)_{ij} + \ldots)P(I + O(\epsilon^k))\varphi_j\,.$$

Since $\langle L_0\varphi_i, \varphi_j^*\rangle = 0$, $(B_0)_{ij} = 0$; so comparing the left and right-hand sides of the above equation we conclude that $(B_1)_{ij} = \ldots = (B_{k-1})_{ij} = 0$ and that

$$PL_kP\varphi_i = PL_k\varphi_i = \sum_{j=1}^n (B_k)_{ij}\varphi_j\,.$$

Operating on this equation with φ_m^* and noting that

$< Pu, \varphi_m > \; = \; < u, P\,\varphi_m > \; = \; < u, \varphi_m >$ we get

$$< L_k \varphi_i, \varphi_m > \; = \; (B_k)_{im}$$

from which (4.10) follows.

We are now ready to prove the main result. In our stability analysis we are led to consider the eigenvalues of the finite dimensional operator $L(\epsilon)E(\epsilon)$. These eigenvalues arise from the perturbation of the critical n-fold eigenvalue at the origin as one moves out along the bifurcation curve from the branch point. The content of the lemma is that, to lowest order, the eigenvalues of $L(\epsilon)E(\epsilon)$ coincide with those of $PL(\epsilon)P$. In the theorem below, we show that $PL(\epsilon)P$ coincides, to lowest order, with the Jacobian of the reduced bifurcation equations.

Theorem 4.3. Consider a general bifurcation problem $G(\lambda, u) = 0$ where $G_u(0,0)$ is a Fredholm operator with an n-dimensional kernel. Let the bifurcation equations be denoted by $F(\lambda, v) = 0$, suppose these have a one-parameter family of solutions

$$\lambda = \epsilon^m \tau_0 \qquad v = \epsilon^\ell \xi \quad , \quad \xi = \xi_0 + \epsilon \xi_1 + \dots ,$$

and suppose that

$$F(\epsilon^m \tau_0, \epsilon^\ell \xi) \; = \; \epsilon^k Q(\tau_0, \xi_0) + O(\epsilon^{k+1}) ,$$

where $k > \max\{m, \ell\}$. The reduced bifurcation equations are therefore

$$Q(\tau_0, \xi_0) = 0.$$

Then the linear finite dimensional operator $B(\epsilon) = L(\epsilon)E(\epsilon)$ given by (4.9) has the form

$$B(\epsilon) = \epsilon^{k-\ell} Q_w(\tau_0, \xi_0) + O(\epsilon^{k-\ell+1}) \tag{4.12}$$

where Q_w denotes the Frechet derivative of the mapping Q. The stability of the bifurcating solutions, being determined by the eigenvalues of $B(\epsilon)$ for small ϵ, is therefore determined in the neighborhood of the bifurcation point by the Jacobian of the reduced bifurcation equations.

Proof. Let $L(\epsilon) = G_u(\lambda(\epsilon), u(\epsilon))$, let $E(\epsilon)$ be the projection constructed in Lemma 4.2, with $E(0) = P$, and let $PL(\epsilon) = O(\epsilon^k)$. Then, according to the lemma, the operators $L(\epsilon)E(\epsilon)$ and $PL(\epsilon)P$ coincide up to order ϵ^k. Therefore, near the branch point, it suffices to examine the spectrum of the operator $PL(\epsilon)P$. The Frechet derivative of the bifurcation equations is

$$F_v(\lambda, v) = PG_u(\lambda, v + \psi(\lambda, v))(I + \psi_v)P.$$

Since ψ_v vanishes at the bifurcation point I dominates the term $I + \psi_v$ as $\epsilon \to 0$, so to lowest order

$$F_v(\lambda, v) \equiv PG_u(\lambda, v + (\psi, v))P \quad \text{(mod lowest order)}.$$

We now introduce the scaling $\lambda = \epsilon^m \tau_0$, $v = \epsilon^\ell \xi$ and put

$$H(\epsilon, \tau_0) = F(\epsilon^m \tau_0, \epsilon^\ell \xi).$$

By hypothesis H has the form

$$H(\epsilon, \tau_0, \xi) = \epsilon^k Q(\tau_0, \xi_0) + O(\epsilon^{k+1}) ;$$

and therefore, to order ϵ^{k+1},

$$H_w = \epsilon^k Q_w(\tau_0, \xi_0) = \epsilon^{\ell} F_v(\epsilon^m \tau_0, \epsilon^{\ell} \xi_0) \qquad \text{(lowest order)}.$$

It follows that, to lowest order terms,

$$PG_u(\lambda(\epsilon), u(\epsilon))P = F_v(\epsilon^m \tau_0, \epsilon^{\ell} \xi_0) = \epsilon^{k-\ell} Q_w(\tau_0, \xi_0) ,$$

which completes the proof of Theorem 4.3.

4. Stability of Bifurcating Solutions at a Simple Eigenvalue.

We may use Theorem 4.3 to give a direct analytical proof of the stability results stated in Theorem 3.1, namely that at a simple eigenvalue supercritical solutions are stable and subcritical solutions are unstable. Recall that the bifurcation equation at a simple eigenvalue takes the form

$$f(\lambda, v) = \sigma'(0)\lambda v - b(\lambda)v^k + \ldots$$

(Cf. 3.6; the quantity $\sigma'(0)$ is the derivative of the critical eigenvalue $\sigma(\lambda)$ as λ crosses the critical value $\lambda_c = 0$.) The reduced equation, obtained from the scaling

$$\lambda = \epsilon^{k-1}\tau \qquad v = \epsilon\xi ,$$

is

$$f(\epsilon^{k-1}\tau, \epsilon\xi) = \epsilon^k Q(\tau, \xi) + O(\epsilon^{k+1})$$

$$Q(\tau, \xi) = \sigma'(0)\tau\xi - b\xi^m = 0 .$$

Taking $\tau = b$ the solutions of the reduced equation are

$$\xi = \begin{cases} \pm\,(\sigma'(0))^{1/m-1} & m \text{ odd} \\ (\sigma'(0))^{1/m-1} & m \text{ even} \end{cases}$$

($\sigma'(0) > 0$ since the zero solution becomes unstable as λ crosses 0 and so $\sigma(\lambda)$ crosses the origin from left to right.) In either case

$$Q_\xi(b, \xi) = b\sigma'(0)(1-m) .$$

According to Theorem 4.3 the leading term in the expansion of the critical eigenvalue is

$$\mu(\epsilon) = \epsilon^{k-1} b\sigma'(0)(1-k)$$

while

$$\lambda(\epsilon) = \epsilon^{k-1} b .$$

Therefore

$$\frac{\mu(\epsilon)}{\lambda(\epsilon)} = \sigma'(0)(1-m) + O(\epsilon)$$

and, for small ϵ, μ and λ are of opposite sign. Therefore $\lambda > 0$ implies $\mu < 0$ and stability, and so forth.

5. Covariance of the Bifurcation Equations

Let equation (1.1) be covariant with respect to a group representation T_g of some group $\mathcal{G}$. That is, assume that

$$T_g G(\lambda, u) = G(\lambda, T_g u) \tag{3.13}$$

where T_g is a representation of $\mathcal{G}$ on the Banach spaces $\mathcal{E}$ and $\mathcal{F}$. Condition (4.13) is a natural one in physical theories. It is a mathematical statement of the axiom that the equations governing a physical system be independent of the observer. For example, the Navier-Stokes equations

$$\begin{aligned} \Delta u_k - \frac{\partial p}{\partial x_k} &= u_j \frac{\partial u_k}{\partial x_j} \\ \frac{\partial u_i}{\partial x_i} &= 0 \end{aligned} \tag{4.14}$$

are covariant with respect to the group of rigid motions. Let $g = \{O,a\}$ be a rigid motion in $\mathbb{R}^3$; that is $g\underline{x} = O\underline{x} + a$ where O is a 3×3 orthogonal transformation and a is a vector in $\mathbb{R}^3$. Equations (4.14) are covariant with respect to the representation

$$T_g \begin{bmatrix} u_1 \\ u_2 \\ u_3 \\ p \end{bmatrix}(x) = \left[\begin{array}{ccc|c} & & & 0 \\ & O & & 0 \\ & & & 0 \\ \hline 0 & 0 & 0 & 1 \end{array}\right] \begin{bmatrix} u_1 \\ u_2 \\ u_3 \\ p \end{bmatrix}(g^{-1}\underline{x}) .$$

Just set $u = (u_1, u_2, u_3, p)$ and

$$G(u) = \begin{pmatrix} \Delta u_1 - \frac{\partial p}{\partial x_1} - u_j \frac{\partial u_1}{\partial x_j} \\ \Delta u_2 - \frac{\partial p}{\partial x_2} - u_j \frac{\partial u_2}{\partial x_j} \\ \Delta u_3 - \frac{\partial p}{\partial x_3} - u_j \frac{\partial u_3}{\partial x_j} \\ \frac{\partial u_i}{\partial x_i} \end{pmatrix}$$

For details see Kirchgässner and Kielhöfer [3].

We now prove

Theorem 4.4. The operator $L_0 = G_u(0,0)$ commutes with T_g and the kernel $\mathcal{N}$ is invariant under T_g. The bifurcation equations $F(\lambda, v)$ are then covariant with respect to the finite dimensional representation T_g restricted to $\mathcal{N}$; that is, $T_g F(\lambda, v) = F(\lambda, T_g v)$.

Proof. We first show that T_g commutes with L_0. Differentiating (4.13) with respect to u, we get

$$T_g G_u(\lambda, u) = G_u(\lambda, T_g u) T_g ;$$

and setting $\lambda = 0$, $u = 0$ we obtain the desired result. Since T_g commutes with L_0 it follows immediately that the kernel $\mathcal{N}$ of L_0 is invariant under T_g. Therefore T_g restricted to $\mathcal{N}$ is a finite dimensional representation of the group $\mathcal{G}$. Now the projection P constructed in Chapter II can be represented by the residue formula

$$P = \frac{1}{2\pi i} \int_C (z - L_0)^{-1} dz$$

where C encloses the origin and no other points in the spectrum of L_0. From this representation for P it is immediate that P and $Q = I - P$ commute with the representation T_g. Applying T_g to (4.1) we have

$$T_g QG(\lambda, v + \psi(\lambda, v)) = QG(\lambda, T_g v + T_g \psi(\lambda, v)) = 0.$$

On the other hand, replacing v by $T_g v$ we get

$$QG(\lambda, T_g v + \psi(\lambda, T_g v)) = 0.$$

Since the solution of (4.1) is unique we must have $T_g \psi(\lambda, v) = \psi(\lambda, T_g v)$. Now applying T_g to (4.2) and using this result we have

$$\begin{aligned} T_g F(\lambda, v) &= T_g PG(\lambda, v + \psi(\lambda, v)) \\ &= PG(\lambda, T_g v + T_g \psi(\lambda, v)) \\ &= PG(\lambda, T_g v + \psi(\lambda, T_g v)) \\ &= F(\lambda, T_g v) \ , \end{aligned}$$

which proves the theorem.

6. Tensor Character of the Bifurcation Equations.

The covariance of the bifurcation equations which was established in Theorem 4.4 can be used to construct the bifurcation equations up to unknown scalar multiples once the group $\mathcal{G}$ and the representation $T_g|_{\eta}$ are known. This results in an enormous reduction in the complexity of a bifurcation problem at a multiple eigenvalue; and it enables one to classify bifurcation problems according to the geometry of the problem and to develop a mathematical theory of bifurcation which is independent of the physical mechanisms of any particular problem. Group representation theory is a linear one, while bifurcation is a non-linear phenomenon. Nevertheless, group representation theory applies in a natural and elegant way due to the tensor character of the branching equations on the one hand and the theory of tensor products of group representations on the other [5]. We shall discuss the elements of group representation theory in the next chapter. For now we discuss the tensor character of the bifurcation equations.

Let V be a vector space and V^* its dual. We choose dual bases for V and V^*, denoted by $\{\varphi_1, \ldots, \varphi_n\}$ and $\{\varphi_1^*, \ldots, \varphi_n^*\}$, and such that $\langle\varphi_i, \varphi_j^*\rangle = \delta_{ij}$. A tensor of type (m, n) over V is an $n+m$ linear functional $F(u_1, \ldots, u_m, v_1^*, \ldots, v_n^*)$ in the variables $u_i \in V$ and v_j^* in V^*. For example, a tensor of type $(2,0)$ is a bilinear functional $B(u_1, u_2)$.

There is a natural way to define the product of two tensors. For example, if φ and ψ are tensors of type $(m,0)$ and $(n,0)$, we define $\varphi \otimes \psi$ to be the tensor given by

$$(\varphi \otimes \psi)(u_1, \ldots, u_m; v_1, \ldots, v_n) = \varphi(u_1, \ldots, u_m)\psi(v_1, \ldots, v_n).$$

The reader may make the modifications suitable to cover the general case. In general, if V and W are finite dimensional vector spaces with bases $\{\underline{v}_i\}$, $\{\underline{w}_i\}$, $i = 1, \ldots, n$, $j = 1, \ldots, m$ then we construct the formal tensor product space $V \otimes W$ which is the mn-dimensional vector space spanned by the basis elements $\{\underline{v}_i \otimes \underline{w}_j\}$. The elements of $V \otimes W$ are thus all of the form

$$\sum_{i,j} \alpha_{ij} \underline{v}_i \otimes \underline{w}_j \ .$$

The tensor product is to obey the following rules

$$\alpha(\underline{v} \otimes \underline{w}) = (\alpha \underline{v}) \otimes \underline{w} = \underline{v} \otimes (\alpha \underline{w})$$

$$(\underline{u} + \underline{v}) \otimes \underline{w} = \underline{u} \otimes \underline{w} + \underline{v} \otimes \underline{w}$$

$$\underline{v} \otimes (\underline{u} + \underline{w}) = \underline{v} \otimes \underline{u} + \underline{v} \otimes \underline{w}$$

For example, consider the functionals φ^* and ψ^* in V^*. We define the tensor product $\varphi^* \otimes \psi^*$ to be the $(2,0)$ tensor

$$(\varphi^* \otimes \psi^*)(\underline{u}, \underline{v}) = \varphi^*(\underline{u})\psi^*(\underline{v}) \ .$$

This definition of $\otimes$ obeys all the above rules. Likewise, we can identify the pair of vectors $(\underline{u}, \underline{v})$ with the tensor product $\underline{u} \otimes \underline{v}$. By this identification, $(V \otimes V)^* \cong V^* \otimes V^*$, etc.

A linear mapping $A: V \to V$ can be identified with an element in $V \otimes V^*$ in a natural way by forming the bilinear functional $B(\underline{u}, v^*) = \langle Au, v^* \rangle$. Thus with A is associated a tensor of type (1, 1), that is, an element in $V \otimes V^*$. Conversely, given such a tensor $B(\underline{u}, \underline{v}^*)$ we define an operator A by considering the linear functional $\underline{v}^* \to B(\underline{u}, \underline{v}^*)$ for fixed $\underline{u}$. By duality, there must be an element $A(\underline{u})$, depending on $\underline{u}$, such that $B(\underline{u}, \underline{v}^*) = \langle A(\underline{u}), \underline{v}^* \rangle$. Then, since the left side is linear in $\underline{u}$, the right side is also linear in $\underline{u}$. We could also proceed as follows. Represent the linear operator A by

$$\mathbf{A}\underline{u} = \sum_{j=1}^{n} a_j(\underline{u})\varphi_j .$$

By linearity (and the linear independence of the set $\{\varphi_j\}$, the functionals $a_j(\underline{u})$ must be linear in $\underline{u}$. They therefore have the representation

$$a_j(\underline{u}) = \langle \underline{u}, \sum_{k=1}^{n} a_{jk}\varphi_k^* \rangle$$

$$= \sum_{k=1}^{n} a_{jk} \langle \underline{u}, \varphi_k^* \rangle .$$

Therefore

$$A\underline{u} = \sum_{j,k=1}^{n} a_{jk} \langle \underline{u}, \varphi_k^* \rangle \varphi_j .$$

and

$$A = \sum_{j,k=1}^{n} A_{jk}\, \underline{\varphi}_j \otimes \underline{\varphi}_k^*$$

where it is understood that $(\underline{\varphi}_j \otimes \underline{\varphi}_k^*)(\underline{u}) = \langle \underline{u}, \varphi_k^* \rangle \underline{\varphi}_j$.

These same considerations carry over immediately to k-linear operators for arbitrary k. If B is a k-linear operator we consider the (k+1)-linear functional

$$F(\underline{u}_1, \ldots, \underline{u}_k, v^*) = \langle B(\underline{u}_1, \ldots, \underline{u}_k), \underline{v}^* \rangle ;$$

and conversely, such a (k+1)-linear functional F induces a k-linear operator B. There is thus a natural isomorphism between the spaces of k-linear operators on V and the tensor space $V^{\otimes k} \otimes V^*$. In the expansion (4.4) every term $B_k(\lambda; v, \ldots, v)$ in the bifurcation equations may be identified with a tensor in $V^{\otimes k} \otimes V^*$.

Suppose a group $\mathcal{G}$ acts on V through a finite dimensional representation T_g. The k-linear mapping B is covariant if $T_g B(\underline{u}_1, \ldots, \underline{u}_k) = B(T_g \underline{u}_1, \ldots, T_g \underline{u}_k)$. Since the branching equations are covariant, each successive term is also covariant. Furthermore, each mapping B_k can be considered to be symmetric in all its variables since we only consider expressions of the type $B_k(\lambda; v, \ldots, v)$.

The fact that we are interested only in mappings which are symmetric in all their variables permits a simplification of the tensor analysis. The algebra of symmetric tensors over a vector space V is isomorphic to the algebra of polynomials in $z_1, \ldots, z_n$ $(n = \dim V)$. For, consider a tensor product of k vectors

$$\varphi_{i_1} \otimes \cdots \otimes \varphi_{i_k} \quad .$$

We want only the completely symmetric part of this tensor, so we symmetrize this vector as follows

$$\frac{1}{k!} \sum_{\pi \in S_k} \varphi_{i_{\pi(1)}} \otimes \ldots \otimes \varphi_{i_{\pi(k)}} \tag{4.16}$$

where π runs over all permutations of k objects. The tensor product (4.16) is then entirely determined by the "occupation numbers" $\{n_i\}$, where n_j is the number of times φ_j occurs in the product $\varphi_{i_1} \otimes \ldots \otimes \varphi_{i_k}$. The vector (4.16) may thus be identified with the k-nomial

$$z_1^{n_1} \cdots z_n^{n_n} ,$$

and, in general, an arbitrary symmetric tensor of order k can be identified with a homogeneous polynomial of degree k

$$\sum_{|\alpha| = k} A_\alpha z_1^{\alpha_1} \cdots z_n^{\alpha_n} , \quad |\alpha| = \alpha_1 + \ldots + \alpha_n . \tag{4.17}$$

The vector space V is identified with linear polynomials in $z_1, \ldots, z_n$ and the space $(V^{\otimes k})_S$ (the symmetric part of $V^{\otimes k}$) is then isomorphic to polynomials (4.17). The mappings $B(v, \ldots, v)$ which are homogeneous of degree k then take the form

$$\begin{pmatrix} b_1(z_1, \ldots, z_n) \\ \cdot \\ \cdot \\ \cdot \\ b_n(z_1, \ldots, z_n) \end{pmatrix}$$

where each b_j is a homogeneous polynomial of degree k.

In order to construct the covariant bifurcation equations, it is therefore entirely natural to study the group action on the algebra of polynomials in $z_1, \ldots, z_n$. The problem is to construct all homogeneous mappings in the variables $z_1, \ldots, z_n$ once the group action on $z_1, \ldots, z_n$ is known. This will be the topic of the next three chapeters.

References

1. N. Dunford and J. Schwartz, Linear Operator Theory, Vol. I Interscience, New York.

2. L. Graves, "Remarks on singular points of functional equations," Trans. Amer. Math. Soc. 79 (1955), 150-157.

3. H. Kielhofer and K. Kirchgassner, "Stability and bifurcation in fluid mechanics," Rocky Mountain Jour. Math. 3 (1973), 275-318.

4. D. Sather, "Branching of solutions of nonlinear equations," Rocky Mountain Jour. Math. 3 (1973), 203-250.

5. D. H. Sattinger, "Group representation theory and branch points of nonlinear functional equations," SIAM Jour. Math. Anal. 8 (1977), 179-201.

6. D. H. Sattinger, "Group representation theory, bifurcation theory, and pattern formation," Jour. Functional Analysis, 77 (1978),

V

ELEMENTS OF GROUP REPRESENTATION THEORY

1. Representations; Reducibility; Character Tables

A representation of a group $\mathcal{G}$ on a vector space V is a homomorphism $g \to T(g)$ of $\mathcal{G}$ into the group of invertible linear transformations on V:

$$T(g_1 g_2) = T(g_1)T(g_2)$$

When V is an inner product space with Hermitian inner product (u,v), $T(g)$ is called a unitary representation provided $(T(g)u, T(g)v) = (u,v)$ for all u, v, and g. On the other hand, if V is a Banach space, let V^* be the dual space and denote the bilinear pairing by $< u, \varphi^* >$. If L is any linear transformation on V denote its adjoint by L^+: $< Lu, \varphi^* > = < u, L^+ \varphi^* >$. The contragradient representation on V^* is defined to be $\tilde{T}(g) = T^+(g^{-1})$, and it has the property that

$$< T(g)u, \tilde{T}(g)\varphi^* > = < u, \varphi^* > \tag{5.1}$$

A subspace V' is invariant under $T(g)$ if $T(g)V' \subseteq V'$ for all $g \in \mathcal{G}$; and $T(g)$ is said to be <u>reducible</u> if it possesses a proper invariant subspace. Otherwise T is said to be <u>irreducible</u>. Two representations T_1 and T_2 are equivalent if there exists a nonsingular linear transformation S such that

$$T_1(g) = ST_2(g)S^{-1}$$

for all $g \in \mathcal{G}$. If T_1 and T_2 are two representations of $\mathcal{G}$ on vector spaces V and W respectively, we construct their direct sum $T_1 \oplus T_2$ on $V \oplus W$ by

$$(T_1 \oplus T_2)(u, v) = (T_1 u, T_2 v).$$

The matrix of $T_1 \oplus T_2$ on $V \oplus W$ therefore takes the form

$$\left(\begin{array}{c|c} T_1 & 0 \\ \hline 0 & T_2 \end{array}\right)$$

Conversely, as we shall see below, every finite-dimensional representation of a finite or compact group $\mathcal{G}$ can be decomposed into a direct sum of irreducible representations.

A group $\mathcal{G}$ is of finite order $N = |\mathcal{G}|$ if it has N elements in it. Examples of finite groups are the permutation groups S_n and the symmetry groups of polygons and polyhedrons. It is convenient to use the cycle notation. The permutation

$$\begin{pmatrix} 1 & 2 & 3 & 4 & 5 & 6 & 7 & 8 \\ 2 & 4 & 6 & 8 & 7 & 5 & 3 & 1 \end{pmatrix},$$

for example, is represented in cycle notation by (1248)(3657). Two cycles are multiplied together by starting from the right. For example, if $\alpha = (123456)$ and $\beta = (35)(26)$ then $\alpha\beta = (12)(36)(45)$. According to Cayley's theorem, every finite group is isomorphic to a subgroup of a permutation group. The cycles α and β, for example, are the generators of the symmetry group of the hexagon

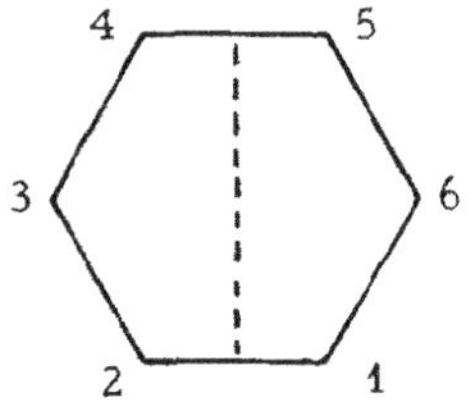

α is the rotation through 60° and β is the reflection across the axis joining the vertices 1 and 4; their product $\alpha\beta$ is the reflection across the median indicated by a dotted line in the diagram.

A continuous transformation group $\mathcal{G}$ is a k-parameter Lie group if locally it has the differential structure of an analytic k-dimensional manifold in which the group operation $(g, h) \rightarrow g^{-1}h$ is differentiable. The Lie algebra of $\mathcal{G}$, denoted by $\mathfrak{g}$, is the tangent space at the identity. If $\mathcal{G}$ is compact there exists an invariant measure μ on $\mathcal{G}$ such that

$$\int_{\mathcal{G}} f(hg)d\mu(g) = \int_{\mathcal{G}} f(gh)d\mu(g) = \int_{\mathcal{G}} f(g)d\mu(g)$$

for any continuous function f on $\mathcal{G}$. We always assume the measure is normalized so that $\int d\mu = 1$.

Theorem 5.1. Every representation T of a finite or compact group $\mathcal{G}$ which acts on a Hilbert space $\mathcal{H}$ is equivalent to a unitary representation.

Proof. (See Wigner [7]). We construct a Hermitian operator

$$H = \sum_g T_g^* T_g \quad \text{or} \quad \int_{\mathcal{G}} T_g^* T_g \, d\mu(g)$$

in the case of a finite or compact group $\mathcal{G}$ respectively. H is positive definite since

$$(Hu,u) = \sum_g (T_g^* T_g u,u) = \sum_g (T_g u, T_g u) > 0.$$

Let H be diagonalized by U and put $d = U^{-1}HU$. Then d is a positive definite symmetric operator with positive square root $d^{1/2}$. Let $\overline{T}(g) = U^{-1}T(g)U$ and $\overline{\overline{T}}(g) = d^{1/2}\overline{T}(g)d^{-1/2}$. Then

$$d = \sum_g \overline{T}(g^*)\overline{T}(g) \quad \text{and} \quad I = d^{-1/2}\sum_g \overline{T}_g^* \overline{T}_g d^{-1/2} .$$

Therefore

$$\begin{aligned}
\overline{\overline{T}}^*(g)\overline{\overline{T}}(g) &= d^{-1/2}\overline{T}^* d^{1/2} d^{1/2}\overline{T} d^{-1/2} \\
&= d^{-1/2}\overline{T}^* d^{1/2}(d^{-1/2}\sum_h \overline{T}^* \overline{T}_h d^{-1/2}) d^{1/2}\overline{T} d^{-1/2} \\
&= d^{-1/2}\sum_h \overline{T}^*(g)\overline{T}^*(h)\overline{T}(h)\overline{T}(g) d^{-1/2} \\
&= d^{-1/2}\sum_h (\overline{T}(hg))^*(\overline{T}(hg))\, d^{-1/2} \\
&= I ,
\end{aligned}$$

which proves that $\overline{\overline{T}}(g)$ is a unitary representation.

Therefore, for compact or finite groups acting on a Hilbert space $\mathcal{H}$ we may (and shall) always assume the representation is unitary. Furthermore, whenever the dimension of the representation is finite we may always assume the vector space has a Hilbert space structure. This enables us to apply the standard results about finite-dimensional representations, which are described below.

Theorem 5.2 (Schur's lemma). Let $T(g)$ be a finite-dimensional representation which commutes with a linear operator M:

$$MT(g) = T(g)M \qquad \text{for all } g \in \mathcal{G}.$$

If T is irreducible, then M is a scalar multiple of the identity. On the other hand, if the only M which commute with $T(g)$ are scalar multiples of the identity, then $T(g)$ is irreducible.

Schur's lemma is the cornerstone of group representation theory and will be used directly in our work on bifurcation theory.

Theorem 5.3. Let $T_{ij}^{(1)}$ and $T_{k\ell}^{(2)}$ be unitary irreducible matrix representations of a finite group $\mathcal{G}$. Then

$$\frac{1}{N}\sum_{g\in} \sqrt{\ell_\mu}\, T_{ij}^{(\mu)}(g)\sqrt{\ell_\nu}\, \overline{T_{k\ell}^{(\nu)}(g)} = \delta_{ik}\delta_{j\ell}\delta_{\mu\nu} \tag{5.2}$$

where ℓ_μ is the dimension of the representation $T^{(\mu)}$, and N is the order of $\mathcal{G}$. The same result holds in the case of a compact continuous group $\mathcal{G}$ with the sum

$$\frac{1}{N} \sum_{\mathcal{G}} \quad \underline{\text{replaced by}} \quad \int_{\mathcal{G}} d\mu(g) .$$

The vectors $\sqrt{\ell_\mu}\, T_{ij}^{(\mu)}(g)$ form a complete orthonormal system of functions on $\mathcal{G}$

If $\mathcal{G}$ is a compact group the completeness of the functions $T_{ij}^{(\mu)}(g)$ is known as the Peter-Weyl theorem.

Two elements of $\mathcal{G}$, say g and h, are conjugate if there is an element P such that $g = php^{-1}$. Conjugacy is an equivalence relations which partitions $\mathcal{G}$ into separate classes, called conjugacy classes.

Theorem 5.4. For finite groups $\mathcal{G}$ the number of inequivalent irreducible representations is equal to the number c of conjugacy classes, and

$$\ell_1^2 + \ldots + \ell_c^2 = N .$$

The last relation is simply a reflection of the fact that the vectors $\sqrt{\ell_\mu}\, T_{ij}^{(\mu)}(g)$ form a complete set over $\mathcal{G}$.

2. Characters; Character Tables

The character of a representation $T(g)$ is the trace

$$\chi(g) = \mathrm{Tr}\, T(g) ;$$

the character is well-defined independently of the matrix representation since the trace is an invariant. The characters of $\mathcal{G}$ are the characters of the inequivalent irreducible representations of $\mathcal{G}$. If g and h are conjugate, then

$$\chi(g) = \mathrm{Tr}\, T(php^{-1}) = \mathrm{Tr}\, T(p)T(h)T(p^{-1})$$

$$= \mathrm{Tr}\, T(h) = \chi(h) ,$$

so the characters are constant on conjugacy classes; they are therefore called class functions, since their values depend only on the conjugacy class. The characters form an orthonormal system of functions on the conjugacy classes.

Theorem 5.5. The characters of the irreducible representations satisfy the following orthonormality conditions

$$\frac{1}{N} \sum_{g \in \mathcal{G}} \chi^{\mu}(g)\, \overline{\chi^{\nu}(g)} = \delta_{\mu\nu} \tag{5.3}$$

The same relation holds in the case of a compact continuous group provided the sum is replaced by integration with respect to the normalized invariant measure.

Equation (5.3) is obtained by setting $i = j$ and $k = \ell$ in (5.2) and summing over $i = 1$ to ℓ_{μ}, $j = 1$ to ℓ_{ν}. Since the characters are constant on conjugacy classes we may rewrite (5.3) as

$$\frac{1}{N} \sum_{i=1}^{c} m_i \chi_i^{\mu}\, \overline{\chi_i^{\nu}} = \delta_{\mu\nu} . \tag{5.4}$$

where χ_i^μ is the value of the character χ^μ on the i-th class and m_i is the number of elements in the i-th class. It is convenient to display the values of the characters of a finite group in a character table. We enumerate the conjugacy classes along the top row and the inequivalent representations down the left column.

Example 5.6. Let us discuss the characters and representations of the symmetry group D_3 of the equilateral triangle

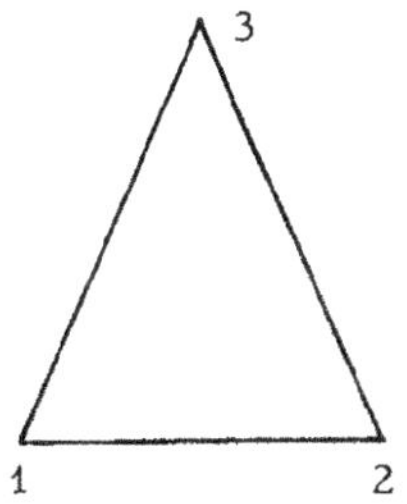

The generators of this group are $g = (123)$ and $h = (12)$ and the conjugacy classes are $\{e\}$, $\{g, g^2\}$, and $\{h, gh, g^2h\}$. There are therefore three irreducible representations of dimensions n_1, n_2, and n_3. According to Theorem 5.4 these must satisfy the condition

$$n_1^2 + n_2^2 + n_3^2 = 6 ;$$

and if we order them so that $n_1 \leq n_2 \leq n_3$ the only possible solution is $n_1 = 1$, $n_2 = 1$, and $n_3 = 2$. Accordingly there are two one-dimensional representations and one two-dimensional representation.

One obvious representation is the identity representation which assigns 1 to every group element. This scalar representation is then its own character, and $\chi^{(1)}(\sigma) = 1$ for all $\sigma \in D_3$. A second representation is obtained by assigning the value $+1$ to pure rotations and -1 to rotation-reflections. Finally, a two-dimensional representation is obtained by considering the standard action of D_3 as rotations and reflections in the plane:

$$T^{(3)}(g) = \begin{pmatrix} -\frac{1}{2} & \frac{\sqrt{3}}{2} \\ -\frac{\sqrt{3}}{2} & -\frac{1}{2} \end{pmatrix}, \qquad T^{(3)}(h) = \begin{pmatrix} -1 & 0 \\ 0 & 1 \end{pmatrix}.$$

Here $T^{(3)}(g)$ is the matrix for the rotation through 120° and $T^{(3)}(h)$ is the matrix of the reflection in the y-axis. For any representation $\chi(e) = \mathrm{Tr}\, I = n_j$, where n_j is the dimension of the given representation. It is customary to write the identity representation along the top row and the identity element (which is always in a conjugacy class by itself) in the first column. Consequently the character table always has one's along the top row and $n_1, n_2, \ldots$ down the first column.

The character table for D_3 is given below.

D_3	$\{e\}$	$\{g, g^2\}$	$\{h, gh, g^2h\}$
$\chi^{(1)}$	1	1	1
$\chi^{(2)}$	1	1	-1
$\chi^{(3)}$	2	-1	0

Note that the orthogonality conditions (5.4) are satisfied.

Theorem 5.7. A finite-dimensional representation T of a finite or compact group can be decomposed into a direct sum of irreducible representations. The decomposition may be written as

$$T = \sum_{j=1}^{c} a_j T^{(j)} \quad ,$$

where a_j is the number of times $T^{(j)}$ occurs in T. The multiplicities a_j are uniquely determined by the formula

$$a_j = \frac{1}{N} \sum_{g \in} \chi(g) \overline{\chi^{(j)}(g)} \tag{5.5}$$

where $\chi(g) = \mathrm{Tr}\, T(g)$ and $\chi^{(j)}$ is the character of the j-th irreducible representation.

Proof. We may assume, by Theorem 5.1, that T is unitary. Consequently, if T leaves a subspace M invariant, it also leaves its orthogonal complement $M^{\perp}$ invariant; for if $u \in M^{\perp}$ and $v \in M$, then

$$(T(g)u, v) = (u, T^{*}(g)v) = (u, T^{-1}(g)v)$$
$$= (u, T(g^{-1})v) = 0 \quad ,$$

since $T(g^{-1})M \subseteq M$, and it follows that $T(g)u \in M^{\perp}$ as well. Therefore the vector space V on which T acts decomposes into the direct sum of two invariant subspaces $V = M \oplus M^{\perp}$. Then T restricted to M and $M^{\perp}$ may be considered anew, and the argument repeated. We proceed in this fashion until V is decomposed into the direct sum of irreducible invariant subspaces.

The decomposition of T in this fashion is unique up to order. For, let χ be the character of T. If T is decomposed into a_1 irreducible representations $T^{(1)}$, a_2 representations $T^{(2)}$, etc., then we must have

$$\chi(g) = \sum_{j=1}^{c} a_j \chi^{(j)}(g) \quad .$$

From the orthogonality conditions (5.3), however, we obtain (5.5). Accordingly, the multiplicities a_j are uniquely determined by the character of the representation, and so the uniqueness of the decomposition follows.

3. Tensor Products of Group Representations

Let S and T be representations of a group $\mathcal{G}$ on the vector spaces V and W respectively. We construct the tensor product representation $S \otimes T$ on $V \otimes W$ by defining

$$(S(g) \otimes T(g))(\underline{v} \otimes \underline{w}) = S(g)\underline{v} \otimes T(g)\underline{w}$$

on elements of the type $\underline{v} \otimes \underline{w}$ and extending to the full space $V \otimes W$ by linearity. If B is a tensor of type (k,ℓ) over V and T is a representation on V we may define

$$T^{\otimes k} \otimes \tilde{T}^{\otimes \ell} B(\underline{u}_1, \ldots, \underline{u}_k; \underline{v}_1^*, \ldots, \underline{v}_\ell^*) = B(T(g^{-1})\underline{u}_1, \ldots; \tilde{T}(g^{-1})\underline{v}_1^*, \ldots)$$

A tensor B of type (k, ℓ) is an invariant of $T^{\otimes k} \otimes \tilde{T}^{\otimes \ell} B = B$, that is, if

$$B(T(g^{-1})\underline{u}_1, \ldots, T(g^{-1})\underline{u}_\ell; \tilde{T}(g^{-1})\underline{v}_1^*, \ldots, \tilde{T}(g^{-1})\underline{v}_\ell^*) = B(\underline{u}_1, \ldots; \underline{v}_1, \ldots)$$

for all $\underline{u}_1, \ldots, \underline{u}_\ell \in V$ and all $g \in \mathcal{G}$. If B is a covariant k-linear mapping then the associated tensor

$$F(\underline{u}_1, \ldots, \underline{u}_k; \underline{v}^*) = \langle B(\underline{u}_1, \ldots, \underline{u}_k), \underline{v}^* \rangle$$

is an invariant of the representation $T^{\otimes k} \otimes \tilde{T}$, where $\tilde{T}$ is the contragradient representation. If T happens to be a unitary representation on the inner product space V, then B is a covariant k-k-linear mapping if and only if $F(\underline{u}_1, \ldots, \underline{u}_k, \underline{v}) = (B(\underline{u}_1, \ldots, \underline{u}_k), v)$ is an invariant of $T^{\otimes(k+1)}$.

Lemma 5.8. *Let T be any representation of a finite or compact group $\mathcal{G}$ over a vector space V. Then the number of invariants of T — that is, the number of vectors $\underline{v} \in V$ which are fixed under T(g) for all $g \in \mathcal{G}$ — is equal to*

$$a_1 = \frac{1}{|\mathcal{G}|} \sum_{g \in \mathcal{G}} \chi(g) \qquad (5.6)$$

where χ *is the character of* T.

Proof. An invariant vector is simply a one-dimensional subspace of V that is left fixed by the representation T. Accordingly, the invariants are the subspaces which transform according to the identity representation; and the result of Lemma 5. 8 follows from (5.5).

Lemma 5.9. *The character of the tensor product of two representations is the product of their characters.*

Proof. Let $\{v_j\}$ and $\{w_\ell\}$ be bases for V and W respectively, and let the matrices of S and T relative to these bases be S_{ij} and $T_{k\ell}$. The matrix of $S \otimes T$ relative to the bases $\{v_j \otimes w_\ell\}$ of $V \otimes W$ is then

$$\begin{aligned} (S \otimes T)\underline{v}_i \otimes \underline{w}_k &= S\underline{v}_i \otimes T\underline{w}_k \\ &= \Big(\sum_j S_{ij}\underline{v}_j\Big) \otimes \Big(\sum_\ell T_{k\ell}\underline{w}_\ell\Big) \\ &= \sum_{j,\ell} S_{ij} T_{k\ell} \underline{v}_j \otimes \underline{w}_\ell \ . \end{aligned}$$

The diagonal elements are thus of the form $S_{ii}T_{kk}$. The trace of $S \times T$ is obtained by summing over the diagonal elements:

$$\chi_{S\otimes T} = \operatorname{Tr} S\otimes T = \sum_{i,k} S_{ii}T_{kk} = (\sum_i S_{ii})(\sum_k T_{kk})$$

$$= (\operatorname{Tr} S)(\operatorname{Tr} T) = \chi_S\chi_T .$$

A tensor $B(\underline{u}_1, \ldots, \underline{u}_k)$ in $(V^*)^{\otimes k}$ is symmetric if it is invariant under arbitrary permutations of its arguments. If B is symmetric it is clear that $T^{\otimes k}B$ is also symmetric. Therefore the tensor product representation $T^{\otimes k}$ is reduced by the subspace of symmetric tensors $(V^{\otimes k})_S$. The restriction of $T^{\otimes k}$ to this subspace is denoted by $(T^{\otimes k})_S$. From Lemma 5.8 the character of $T^{\otimes k}$ is χ^k, where χ is the character of T; but the calculation of the character of $(T^{\otimes k})_S$, which we denote by $\chi_{(k)}$, is more complicated. Its computation will be carried out below.

Theorem 5.10. Let T be a representation of a compact or finite group $\mathcal{G}$ on the vector space V and let $c_k(T, \mathcal{G})$ denote the number of symmetric k-linear mappings B which are covariant with respect to T. The coefficients c_k may be computed from the following generating function

$$\sum_{k=0}^{\infty} c_k(T, \mathcal{G})\, z^k = \frac{1}{|\mathcal{G}|} \sum_{g\in\mathcal{G}} \det(I - zT(g))^{-1}\overline{\chi}(g), \tag{5.7}$$

<u>where</u> χ <u>is the character of</u> T <u>and</u> $c_0(T, \mathcal{G}) = 1$. <u>In the case of a continuous group the sum is replaced by integration over the group with respect to the normalized invariant measure.</u>

The coefficients c_k can also be determined from the character χ of T by the following formula

$$c_k(T, \mathcal{G}) = \frac{1}{|\mathcal{G}|} \sum_{g \in \mathcal{G}} \chi_{(k)}(g) \overline{\chi}(g)$$

where

$$\chi_{(k)}(g) = \sum_{\sum_{\ell=1}^{k} \ell i_\ell = k} \frac{\chi^{i_1}(g) \ldots \chi^{i_k}(g^k)}{1^{i_1} i_1!\, 2^{i_2} i_2! \cdots k^{i_k} i_k!} . \qquad (5.8)$$

<u>Proof.</u> In order to count the number of k-linear covariant symmetric mappings we recall that such mappings are in one-to-one correspondence with the invariants of the representation $(T^{\otimes k})_S \otimes \widetilde{T}$ and apply Lemma 5.7. Accordingly, we need to compute the quantity

$$\frac{1}{|\mathcal{G}|} \sum_{g \in \mathcal{G}} \chi_{(k)}(g) \overline{\chi}(g)$$

since $\overline{\chi}$ is the character of $\widetilde{T}$. (In fact, $\mathrm{Tr}\, \widetilde{T}(g) = \mathrm{Tr}\, T^+(g^{-1}) = \mathrm{Tr}\, T(g^{+1}) = \mathrm{Tr}\, T^{-1}(g)$. Since every representation of a compact or finite group is equivalent to a unitary representation, we can assume T is unitary; and then $T^{-1}(g) = T^{*}(g)$, so $\mathrm{Tr}\, T^{-1}(g) = \mathrm{Tr}\, \overline{T^{+}(g)} = \mathrm{Tr}\, \overline{T(g)} = \overline{\chi(g)}$.)

Now let us derive a generating function for the functions $\chi_{(k)}(g)$. Fix the group element g and let the eigenvectors of T be $\underline{e}_1, \dots, \underline{e}_n$ with eigenvalues $\lambda_1, \dots, \lambda_n$. The vector space $(V^{\otimes k})_S$ is spanned by the vectors

$$(\underline{e}_{i_1} \otimes \dots \otimes \underline{e}_{i_k})_S = \sum_{\pi \in S_k} \underline{e}_{i_{\pi(1)}} \otimes \dots \otimes \underline{e}_{i_{\pi(k)}} .$$

Let us denote such a symmetrized vector by the symbol $|m_1 \dots m_n>$ where $m_1 + \dots + m_n = k$. The action of $(T^{\otimes k})_S$ on $|m_1 \dots m_n>$ is

$$(T^{\otimes k})_S |m_1 \dots m_n> = \lambda_1^{m_1} \dots \lambda_n^{m_n} |m_1 \dots m_n>$$

and therefore

$$\chi_{(k)}(g) = \mathrm{Tr}(T^{\otimes k})_S = \sum_{m_1 + \dots + m_n = k} \lambda_1^{m_1} \dots \lambda_n^{m_n} .$$

Multiplying by z^k and summing we get

$$\sum_{k=0}^{\infty} z^k \chi_{(k)}(g) = \sum_{k=0}^{\infty} \sum_{m_1 + \dots + m_n = k} (z\lambda_1)^{m_1} \dots (z\lambda_n)^{m_n}$$

$$= \sum_{m_1, \dots, m_n = 0} (z\lambda_1)^{m_1} \dots (z\lambda_n)^{m_n}$$

$$= \prod_{i=1}^{n} \frac{1}{1 - z\lambda_i} = \det(I - zT(g))^{-1} .$$

To obtain (5.8) proceed as follows. From the relations

$$\det A = e^{\operatorname{Tr}\log A}$$

and

$$\log(1-z)^{-1} = \sum_{n=1}^{\infty} \frac{z^n}{n}$$

we get

$$\det(I - zT(g))^{-1} = e^{\operatorname{Tr}\left(\sum_{k=1}^{\infty} \frac{z^k T(g^k)}{k}\right)}$$

$$= e^{\sum_{k=1}^{\infty} \frac{z^k \chi(g^k)}{k}}$$

$$= \sum_{n=0}^{\infty} \frac{1}{n!}\left(\sum_{k=1}^{\infty} \frac{z^k \chi(g^k)}{k}\right)^n .$$

From the identity

$$\left(\sum_{k=0}^{\infty} a_k z^k\right)^n = \sum_{m=0}^{\infty} z^m \sum_{r_1+\dots+r_n=m} a_{r_1}\cdots a_{r_m}$$

$$= \sum_{m=0}^{\infty} z^m \sum_{\sum_{\ell=1}^{m} \ell i_\ell} a_1^{i_1} a_2^{i_2} \cdots a_m^{i_m} \frac{m!}{i_1!\, i_2! \cdots i_m!}$$

we get

$$\sum_{n=0}^{\infty} \chi_{(n)}(g) z^n = \det(I - zT(g))^{-1}$$

$$= \sum_{n=0}^{\infty} z^n \left(\sum_{\sum_{\ell=1}^{n} \ell i_\ell} \frac{\chi^{i_1}(g)}{1^{i_1} i_1!} \frac{\chi^{i_2}(g^2)}{2^{i_2} i_2!} \cdots \frac{\chi^{i_n}(g^n)}{n^{i_n} i_n!}\right) .$$

Comparing coefficients of z^n we get (5.8). (N.B. These results are closely related to the classical results for the Molien function. Our proof follows that in Jaric and Birman [2].)

Consider the representation $T^{(1)} \oplus T^{(2)} \oplus T^{(3)}$ of the group D_3. Let us compute the number of covariant mappings of degree 2. From (5.8)

$$\chi_{(2)}(g) = \sum_{i_1+2i_2=2} \frac{\chi^{i_1}(g)\,\chi^{i_2}(g^2)}{1^{i_1} i_1!\, 2^{i_2} i_2!}$$

$$= \frac{\chi^2(g)}{2!} + \frac{\chi(g^2)}{2}$$

$$= \frac{1}{2}(\chi^2(g) + \chi(g^2)) \; .$$

The values of the characters χ and $\chi_{(2)}$ are given below

D_3	$\{e\}$	$\{g, g^2\}$	$\{h, gh, g^2h\}$
χ	4	1	0
$\chi_{(2)}$	10	1	2

The number of covariant mappings is

$$\frac{1}{6} \sum_{\sigma \in D_3} \chi_2(\sigma)\,\chi(\sigma) = \frac{1}{6}(40 + 2 + 0) = 7 \; .$$

We shall construct explicitly the 7 bilinear covariant mappings in Example 5.12 below.

Theorem 5.11. Let the bifurcation equations $F_j(z_1, \ldots, z_n)$ be covariant with respect to a representation T. If T is irreducible the linear terms have the form $F_j = \lambda z_j$; while if T is reducible the linear term on each irreducible block is a scalar multiple of the identity, with independent scalar values for the different blocks.

Proof. This result is a direct consequence of Schur's theorem. For if the linear term F_j is of the form $F_j = \sum_k a_{jk} z_k$, then the matrix $A = (a_{jk})$ must commute with the matrix representation T. If T is irreducible then A is a scalar multiple of the identity, by Schur's theorem. When T is reducible A and T still possess the same invariant subspaces, since they commute. Restricting T and A to an invariant irreducible subspace of T we see that A satisfies the conditions of Schur's Theorem there, so A restricted to such invariant irreducible subspaces of T is a scalar multiple of the identity.

4. Computation Methods for Covariant Mappings

Consider the group S^1 — rotations in the plane. This group has an infinite number of real two-dimensional representations

$$R_n(\theta) = \begin{pmatrix} \cos n\theta & \sin n\theta \\ -\sin n\theta & \cos n\theta \end{pmatrix}$$

together with the reflection

$$H = \begin{pmatrix} 1 & 0 \\ 0 & -1 \end{pmatrix}$$

These are the matrix representations relative to the basis $\underline{e}_1 = \binom{1}{0}$ and $\underline{e}_2 = \binom{0}{1}$. We now choose a complex basis

$$\underline{f}_1 = \frac{\underline{e}_1 + i\underline{e}_2}{\sqrt{2}} \qquad \underline{f}_2 = \overline{\underline{f}_1} = \frac{\underline{e}_1 - i\underline{e}_2}{\sqrt{2}} .$$

The unitary matrix

$$U = \begin{pmatrix} \frac{1}{\sqrt{2}} & \frac{i}{\sqrt{2}} \\ \frac{1}{\sqrt{2}} & \frac{-i}{\sqrt{2}} \end{pmatrix}$$

then diagonalizes $R_n(\theta)$:

$$T_n(\theta) = UR_n(\theta)U^* = \begin{pmatrix} e^{-in\theta} & 0 \\ 0 & e^{in\theta} \end{pmatrix}$$

$$K = UHU^* = \begin{pmatrix} 0 & 1 \\ 1 & 0 \end{pmatrix} .$$

The one-dimensional representations $\{e^{in\theta}\}$ are irreducible (obviously) and constitute the characters of the group. The invariant measure is simply $\frac{d\theta}{2\pi}$, so the orthogonality conditions for the characters are the well-known orthogonality conditions for the Fourier series. The Peter-Weyl theorem, which guarantees the completeness of the characters, is then equivalent to the completeness theorem for the Fourier series.

If the original vector space is real then a vector

$$w = z_1 \underline{f}_1 + z_2 \underline{f}_2$$

is real if and only if $z_1 = \bar{z}_2$, so we may use coordinates z and $\bar{z}$. The group action of S_1 on z and $\bar{z}$ is

$$Kz = \bar{z} \qquad K\bar{z} = z$$

$$T_n(\theta)z = e^{in\theta} z \quad , \quad T_n(\theta)\bar{z} = e^{-in\theta} \bar{z} .$$

Let us compute the covariant mappings for T_n. If $F = (F_1, F_2)$ is a covariant mapping we have $T_n F = F T_n$ implies

$$e^{in\theta} F_1(z, \bar{z}) = F_1(e^{in\theta} z, e^{-in\theta} \bar{z}) \tag{5.9a}$$

$$e^{-in\theta} (F_2(z, \bar{z}) = F_2(e^{in\theta} z, e^{-in\theta} \bar{z}) \tag{t. 9b}$$

and $KF = FK$ implies

$$F_1(z, \bar{z}) = F_2(\bar{z}, z). \tag{5. 9c}$$

Also, if F is a real mapping then

$$\overline{F_2(z, \bar{z})} = F_1(z, \bar{z}). \tag{5. 9d}$$

Equations (5.9c) and (5.9d) together imply

$$\overline{F_1(z, \bar{z})} = F_2(z, \bar{z}) = F_1(\bar{z}, z) \tag{5.10}$$

It suffices to compute F_1; expanding F_1 in a power series in z and $\bar{z}$,

$$F_1(z, \bar{z}) = \sum_{r, s} A_{rs} z^r \bar{z}^s$$

we see from (5.10) that the coefficients A_{rs} are real. From (5.9a)

$$e^{in\theta} \sum_{r, s} A_{rs} z^r \bar{z}^s = \sum_{r, s} A_{rs} e^{in(r-s)\theta} z^r \bar{z}^s .$$

This relationship holds for all θ, z and $\bar{z}$ if and only if

$$e^{in\theta} = e^{in(r-s)\theta} ,$$

hence $r-s = 1$. Thus F is covariant with respect to T_n if

$$F_1(z,\bar{z}) = \sum_{k=0}^{\infty} A_k z^{k+1} \bar{z}^k$$

$$= \sum_{k=0}^{\infty} A_k |z|^{2k} z$$

$$= g(|z|^2) z$$

and

$$F_2(z,\bar{z}) = g(|z|^2) \bar{z} .$$

Example 5.12. Now let us calculate the covariant mappings for the two-dimensional representation of D_3. The two-dimensional representation is generated by

$$T(\tfrac{2\pi}{3}) = \begin{pmatrix} e^{2\pi i/3} & 0 \\ 0 & e^{-2\pi i/3} \end{pmatrix}$$

$$K = \begin{pmatrix} 0 & 1 \\ 1 & 0 \end{pmatrix}$$

If the map is real then (5.9c) and (5.10) continue to hold, so it is again only necessary to compute F_1. For this we get

$$\sum_{r,s} A_{rs} e^{2\pi i/3} z^r \bar{z}^s = \sum_{r,s} A_{rs} e^{2\pi i(r-s)/3}$$

and so we must have

$$e^{2\pi i/3} = e^{2\pi i(r-s)/3} ,$$

$$\frac{2\pi}{3}(1-r+s) = 2\pi n \quad , \quad n = 0, \pm 1, \pm 2, \ldots .$$

Therefore

$$1-r+s = 3n \quad , \quad n = 0, \pm 1, \pm 2, \ldots .$$

We can write this condition in the form

$$s = r + (3n-1) \qquad n = 1, 2, \ldots$$

$$r = s + (1-3n) \qquad n = 0, 1, 2, \ldots$$

So the general mapping is

$$\sum_{n=1}^{\infty} g_n(|z|^2)\,\overline{z}^{(3n-1)} + \sum_{n=0}^{\infty} h_n(|z|^2)\,z^{2n+1}$$

or

$$\sum_{n=0}^{\infty} g_n(|z|^2)\,\overline{z}^{3n+2} + h_n(|z|^2)\,z^{3n+1} \quad ,$$

$$= zh(|z|^2, z^3) + \overline{z}^2 g(|z|^2, \overline{z}^3) \tag{5.11}$$

The lowest order terms are going to be

$$F_1 = A_0 z + A_1 \overline{z}^2 + A_2 |z|^2 z + \ldots$$

$$F_2 = A_0 \overline{z} + A_1 z^2 + A_2 |z|^2 \overline{z} + \ldots$$

There is another way to compute the covariant mappings. The invariants of the operation T are polynomials in $z, \overline{z}$ such that $p(Tz, T\overline{z}) = p(z, \overline{z})$. Clearly if $p(z, \overline{z})$ is an invariant then $zp(z, \overline{z})$ and $\overline{z}^2 p(z, \overline{z})$ are covariant functions. The functions

$|z|^2 = z\bar{z}$, z^3, and $\bar{z}^3$ are clearly invariant, and these form an integrity basis for all invariants: every polynomial $p(z, \bar{z})$ which is an invariant of T is a polynomial in $|z|^2$, z^3 and $\bar{z}^3$. The general T-covariant mapping F_1 has the form

$$F_1(z,\bar{z}) = zA(|z|^2, z^3, \bar{z}^3) + \bar{z}^2 B(|z|^2, z^3, \bar{z}^3).$$

This does not immediately seem to agree with the expression (5.11) until one notices that a term of the form $\bar{z}^{3n} z = |z|^2 \bar{z}^2 \bar{z}^{3(n-1)}$, hence any terms in zA which depend explicitly on $\bar{z}^3$ can be transferred over to $\bar{z}^2 B$ and terms in $\bar{z}^2 B$ which contain z^3 can be transferred to zA.

In principle one can always proceed to construct the covariant mappings in the above way -- that is, by first finding an integrity basis for the invariants and then coupling these to the lowest order covariant mappings. The reality condition (5.10) gives

$$\overline{h(|z|^2, z^3)} = h(|z|^2, \bar{z}^3)$$

$$\overline{g(|z|^2, \bar{z}^3)} = g(|z|^2, z^3)$$

and (5.9c) gives

$$F_2(z,\bar{z}) = \bar{z}h(|z|^2, \bar{z}^3) + z^2 g(|z|^2, \bar{z}^3).$$

<u>Example 5.13.</u> Now let us calculate the covariant mappings for the representation $T = T^{(1)} \oplus T^{(2)} \oplus T^{(3)}$ of D_3 discussed in the Example 5.6. We introduce the variables $x, y, z, \bar{z}$, and identify the vector space V with linear polynomials in these four variables. The group actions on these variables are as follows, with $g = (1\,2\,3)$ and $h = (1\,2)$.

$$gx = hx = x$$

$$gy = y \qquad hy = -y$$

$$gz = e^{2\pi i/3} z \qquad g\bar{z} = e^{-2\pi i/3}\bar{z} \qquad hz = \bar{z} \qquad h\bar{z} = z.$$

The invariants are $x, |z|^2, y^2, z^3$, and $\bar{z}^3$. The mapping F has four components $F = (F_1, F_2, F_3, F_4)$ which transform like x, y, z, and $\bar{z}$ respectively. In particular, F_1 is an invariant, so

$$F_1(x, y, z, \bar{z}) = A(x, y^2, |z|^2, z^3, \bar{z}^3).$$

F_3 must transform as in Example 5.12 above, so it must take the form

$$F_3(x, y, z, \bar{z}) = zA(x, y, |z|^2, z^3) + \bar{z}^2 B(x, y, |z|^2, \bar{z}^3).$$

The symmetry $hF = Fh$ implies

$$F_4(x, y, z, \bar{z}) = F_3(x, -y, \bar{z}, z).$$

The lowest order (linear plus quadratic) terms in the covariant mapping therefore are

$$F_1 = \lambda_1 x + ax^2 + by^2 + c|z|^2 + \dots$$

$$F_2 = y(\lambda_2 + dx) + \dots$$

$$F_3 = \lambda_3 z + exz + fyz + g\bar{z}^2 + \dots$$

$$F_4 = \lambda_4 \bar{z} + ex\bar{z} - fyz + gz^2 + \dots \quad .$$

We have obtained these mappings by writing down all possible linear and quadratic terms which have the appropriate transformation properties. Note that we come up with seven independent possible quadratic terms (labelled by the seven parameters $a, \dots, g$). This conforms with our previous computation by characters that there were precisely seven quadratic covariant terms. In an actual physical problem one would have to know the values of $a, \dots, g$ by direct calculation through the Lyapounov-Schmidt procedure. Such a calculation can be a fairly complicated numerical problem. Instead, one could proceed on a general basis by analyzing the possible bifurcations which might possibly take place as the scalar parameters range over all possible values. In the present case, with seven parameters going, even that is a little awkward, and I use the present example only as an illustration of the techniques one might use in calculating the bifurcation equations. In the applications we discuss in the next chapter there are only one or two parameters.

Example 5.14. Bifurcation in the presence of SO(1) versus O(1).

These two groups are, respectively, the group of pure rotations of the circle and the group of rotations/reflections. The corresponding bifurcation problems are entirely different in character. The first case arises in the Hopf bifurcation theorem since the time dependent equations (3.1) are invariant under time translations $t \to t+\gamma$ but not under the time reflection $t \to -t$. When the problem is reduced to an algebraic one the kernel of the linear operator is a two-dimensional subspace of periodic functions, and the group being considered is $\mathbb{R}/\mathbb{Z} \cong SO(1)$. The bifurcation equations are a system of two equations in the variables z and $\bar{z}$ and they are covariant with respect to the representation

$$R(\theta)z = e^{i\theta}z \quad , \quad R(\theta)\bar{z} = e^{-i\theta}\bar{z} .$$

The equations must also satisfy the reality condition

$$\overline{F_1(z,\bar{z})} = F_2(z,\bar{z})$$

since the bifurcation mapping leaves the subspace of real vectors invariant. The reflection symmetry K, however, is missing, and this has an important consequence.

The covariance condition

$$e^{i\theta}F_1(z,\bar{z}) = F_1(e^{i\theta}z, e^{-i\theta}\bar{z})$$

implies, as before, that F_1 must have the form $F_1(z) = z\Phi(|z|^2)$ and $F_2 = \bar{z}\bar{\Phi}(|z|^2)$. We cannot conclude this time that Φ is real-valued because the reflection symmetry K is lacking. Making the dependence on the parameter λ explicit, the bifurcation equations take the form

$$z\Phi(\lambda,|z|^2) = 0 \ , \quad \overline{\bar{z}\Phi(\lambda, |z|^2)} = 0 \tag{5.12}$$

Clearly it is enough to consider the equation $\Phi(\lambda, |z|^2) = 0$, but this is fully determined if Φ is complex; for we must set both real and imaginary parts of Φ equal to zero, and expect to get a bifurcating curve of solutions. But if an additional parameter ω were somehow introduced we would have

$$\Phi(\lambda,\omega,|z|^2) = \Phi_1(\lambda,\omega, |z|^2) + i\Phi_2(\lambda, \omega, |z|^2) = 0,$$

$$\Phi_1(\lambda,\omega, |z|^2) = 0$$

$$\Phi_2(\lambda,\omega,|z|^2) = 0$$

If the Jacobian $\dfrac{\partial(\Phi_1, \Phi_2)}{\partial(\lambda, \omega)}$ is invertible there exists a unique solution $\lambda = \lambda(|z|^2)$, $\omega = \omega(|z|^2)$. In the Hopf bifurcation theorem the parameter ω is introduced as the frequency of the bifurcating solutions.

On the other hand, the nonlinear problem

$$\Delta u + \lambda u + f(u) = 0, \qquad z^2 + y^2 \leq 1$$

$$u|_{\partial D} = 0$$

is invariant under the group operations $\theta \to \theta + \gamma$ and $\theta \to -\theta$. When bifurcation at a double eigenvalue is considered we arrive at null spaces which transform like $R_n(\theta)$ and K in Example (5.11). The bifurcation equations take the form (5.12) again, but this time the function Φ is real-valued. The reality of Φ is due to the reflection symmetry K (see 5.10). Therefore, in this case, all that is required to solve these bifurcation equations is a condition such that $\Phi_\lambda(0,0) \neq 0$.

5. Lie Groups, Lie Algebras

A k-parameter Lie group is a k-dimensional analytic manifold each of whose points is an element of a group. We have already seen one example -- the circle group S^1. This group has two disconnected components -- the pure rotations and the rotation/reflections. S^1 is a one-parameter group. A second important example of a Lie group is the group of rotations in $\mathbb{R}^3$. The group of pure rotations is a connected 3-parameter group; the full orthogonal group is the group of rotations plus rotation/reflections. The full orthogonal group $O(3)$ is represented by the set of all matrices O such that

$$OO^{+} = I. \tag{5.13}$$

To see that the condition (5.13) defines a manifold, we write it down explicitly

$$\sum_{k=1}^{3} O_{ik}O_{jk} = \delta_{ij} .$$

There are six distinct equations here and nine entries in the matrix O, so barring singularities, we should expect to get a smooth three-dimensional manifold, which is the case.

Another important group in applications is the group of rigid motions in $\mathbb{R}^2$ or $\mathbb{R}^3$. This group consists of the rotation group plus translations. If $g = \{O, \underline{a}\}$ then $g\underline{x} = O\underline{x} + \underline{a}$ where $\underline{a} \in \mathbb{R}^2$ or $\mathbb{R}^3$. The composition law is $\{O, a\}\{O', b\} = \{OO', Ob + a\}$, and the group can be represented by matrices in the following form

$$\{O, a\} = \left(\begin{array}{ccc|c} & & & a_1 \\ & O & & a_2 \\ & & & a_3 \\ \hline 0 & 0 & 0 & 1 \end{array}\right)$$

The reader may check that the composition law is maintained by usual matrix multiplication. The groups of rigid motions are locally compact, but not compact.

In these notes we always deal with linear representations of Lie groups on finite-dimensional vector spaces. The Lie algebra of a Lie group $\mathcal{G}$ is the tangent space at the identity element. For matrix representations the Lie algebra may be computed by differentiating along one-parameter subgroups through the identity. For example, in the case of the circle group S^1 we represent the group elements by

$$R(\theta) = \begin{pmatrix} \cos\theta & \sin\theta \\ -\sin\theta & \cos\theta \end{pmatrix}$$

and differentiate with respect to θ:

$$R'(\theta) = \begin{pmatrix} -\sin\theta & \cos\theta \\ -\cos\theta & -\sin\theta \end{pmatrix}$$

The Lie derivative, the infinitesimal generator of the group, is

$$L = R'(\theta) = \begin{pmatrix} 0 & 1 \\ -1 & 0 \end{pmatrix}$$

The group element $R(\theta)$ can then be recovered by expnentiating L:

$$e^{\theta L} = I + \theta L + \theta^2 \frac{L^2}{2!} + \dots = (1 - \frac{\theta^2}{2!} + \frac{\theta^4}{4!} - \dots)\begin{pmatrix} 1 & 0 \\ 0 & 1 \end{pmatrix}$$

$$+ (\theta - \frac{\theta^3}{3!} + \dots)\begin{pmatrix} 0 & 1 \\ -1 & 0 \end{pmatrix} = R(\theta) .$$

The infinitesimal generators of SO(3), the group of rotations in $\mathbb{R}^3$, are respectively

$$L_1 = \begin{pmatrix} 0 & 0 & 0 \\ 0 & 0 & 1 \\ 0 & -1 & 0 \end{pmatrix} \qquad L_2 = \begin{pmatrix} 0 & 0 & 1 \\ 0 & 0 & 0 \\ -1 & 0 & 0 \end{pmatrix}$$

$$L_3 = \begin{pmatrix} 0 & 1 & 0 \\ -1 & 0 & 0 \\ 0 & 0 & 0 \end{pmatrix}$$

These are the generators of the rotations about the x, y, and z axes respectively, They satisfy the commutation relations

$$[L_i, L_j] = \epsilon_{ijk} L_k \tag{5.14}$$

where ϵ_{ijk} is the completely anti-symmetric tensor.

The invariant integral for SO(3) which has the same value for all elements of a given conjugacy class is

$$\frac{1}{\pi}\int_0^{\pi} (1 - \cos\theta)d\theta \quad .$$

Two elements of SO(3) are conjugate if they represent rotations of equal magnitude (although about different axes).

Sometimes it is convenient to represent the Lie algebra in a different manner, for example as an algebra of differential operators. In the case of the translation group on the real line, $x \to x + t$, we represent the group action on functions as

$$(T_t f)(x) = f(x+t)$$

The infinitesimal generator of this group is the differential operator $\frac{d}{dx}$. In fact,

$$\frac{d}{dt} T_t f(x) \Big|_{t=0} = \lim_{t \to 0} \frac{f(x+t) - f(x)}{t} = \frac{df}{dx} .$$

The group can be recovered by exponentiating the operator $\frac{d}{dx}$, as we shall explain.

Formally, if T_t is a group of operators then the infinitesimal generator of T_t is the linear operator

$$Au = \lim_{t \to 0} \frac{T_t u - u}{t} .$$

Now

$$\frac{dT}{dt} u = \lim_{h \to 0} \frac{T_{t+h} u - T_t u}{h} = \lim_{h \to 0} T_t \left(\frac{T_h u - u}{h} \right) = T_t A u$$

so T formally satisfies the differential equation

$$\frac{dT}{dt} = AT .$$

If T is a matrix representation of the real line then this equation may be solved by linear algebra techniques. Its formal solution is given by

$$T(t) = e^{tA} ;$$

the exponential operator e^{tA} may be defined as a power series

$$e^{tA} = \sum_{n=0}^{\infty} \frac{t^n A^n}{n!}$$

if A is a bounded operator. If A is an unbounded operator, then the questions of analysis are more involved, and we shall not go into them here. But in the case $A = \frac{d}{dx}$ discussed above, the process of exponentiation leads to a partial differential equation, as follows:

$$\frac{d}{dt}(T_t u) = A(T_t u) .$$

Write $(T_t u)(x) = u(x, t)$. Then the above equation becomes

$$\frac{\partial}{\partial t}(T_t u) = \frac{\partial u}{\partial t} = \frac{d}{dx}(T_t u) = \frac{\partial u}{\partial x} ,$$

$$\frac{\partial u}{\partial t} = \frac{\partial u}{\partial x} .$$

Denote $u(x, 0)$ by $f(x)$. The solution to the above partial differential equation is then $u(x, t) = f(x + t)$. Thus
$u(x, t) = T_t u(x, 0) = (T_t f)(x) = f(x + t)$, and $\frac{d}{dx}$ does indeed generate the translation group.

In the case of rotations in the plane we have

$$R(\theta) = \begin{pmatrix} x \\ y \end{pmatrix} = \begin{pmatrix} \cos\theta & \sin\theta \\ -\sin\theta & \cos\theta \end{pmatrix} \begin{pmatrix} x \\ y \end{pmatrix}$$

$$= \begin{pmatrix} x\cos\theta + y\sin\theta \\ -x\sin\theta + y\cos\theta \end{pmatrix}$$

and therefore (recall that $(T_g f)(x) = f(g^{-1}x)$)

$$(R(\theta)f)(x, y) = f(x\cos\theta + y\sin\theta,\ -x\sin\theta + y\cos\theta) ,$$

$$\frac{d}{d\theta} R(\theta)f = y\frac{\partial f}{\partial x} - x\frac{\partial f}{\partial y} .$$

Therefore the infinitesimal generator of the rotations in the plane acting on functions is the partial differential operator

$$L_z = y\frac{\partial}{\partial x} - x\frac{\partial}{\partial y} .$$

The rotations can be recovered by exponentiation -- that is, by solving the initial value problem

$$\frac{\partial u}{\partial t} = y\frac{\partial u}{\partial x} - x\frac{\partial u}{\partial y} .$$

The generators of the rotations in $\mathbb{R}^3$ are

$$L_x = y\frac{\partial}{\partial z} - z\frac{\partial}{\partial y}, \quad L_y = z\frac{\partial}{\partial x} - x\frac{\partial}{\partial z}, \quad L_z = x\frac{\partial}{\partial y} - y\frac{\partial}{\partial x} .$$

and these operators satisfy the same commutation relations as L_1, L_2, and L_3 in (5.14).

The generators for the Lie algebra of the group of rigid motions are L_x, L_y, and L_z above together with the operations $\frac{\partial}{\partial x}$, $\frac{\partial}{\partial y}$, and $\frac{\partial}{\partial z}$.

Now let us investigate the action of Lie derivatives on tensor product spaces. Let $\mathcal{G}$ be a Lie group acting on V through a representation T(g) and consider the action of $\mathcal{G}$ on the tensor product space $V \otimes V$. We differentiate the expression

$$T(g)^{\otimes 2}(\underline{v} \otimes \underline{w}) = T(g)\underline{v} \otimes T(g)\underline{w}$$

along a one-parameter subgroup through the identity. Denoting the corresponding Lie derivative by L, we have

$$\frac{d}{dt} T(g(t))\underline{v} \otimes T(g(t))\underline{w}\Big|_{t=0} = L\underline{v} \otimes \underline{w} + \underline{v} \otimes L\underline{w}$$

($g(0) = e$). Thus the Lie derivative L acts on $V \otimes V$ as a <u>derivation</u>

$$L\underline{v} \otimes \underline{w} = L\underline{v} \otimes \underline{w} + \underline{v} \otimes L\underline{w} .$$

Furthermore, since $T^{\otimes 2}$ leaves $(V^{\otimes 2})_S$ invariant, so do the Lie derivatives L.

The bifurcation equations are comprised of polynomials over a vector space. The kernel $\mathcal{N}$ of $G_u(0,0)$ is identified with the space of linear polynomials in $\{z_1, \ldots, z_n\}$ considered as independent variables. The algebra of symmetric tensors over $\mathcal{N}$ is isomorphic to the algebra of polynomials in $z_1, \ldots, z_n$, which we denote by $K\{z_1, \ldots, z_n\}$. A <u>derivation</u> on this algebra is a linear operator J such that

$$J(\alpha f + \beta g) = \alpha J f + \beta J g$$

$$J(fg) = fJg + (Jf)g$$

where α, β are scalars and f, g are polynomials. The preceding paragraphs are summarized in the following theorem:

<u>Theorem 5.15. Let a Lie group</u> $\mathcal{G}$ <u>act on the polynomial algebra</u> $K[z_1, \ldots, z_n]$. <u>Then the Lie algebra</u> $\mathfrak{g}$ <u>of infinitesimal generators</u> $J_1, \ldots, J_k$ <u>acts as an algebra of derivations on</u> K.

The Lie derivatives possess all the usual algebraic properties of differential operators. In particular, $Jz_i^k = kz^{k-1}Jz_i$. When restricted to the linear space spanned by $\{z_1, \ldots, z_n\}$ each J is a linear operator and so

$$Jz_i = \sum_{j=1}^{n} J_{ij} z_j .$$

When the bifurcation equations are covariant with respect to a Lie group, the various terms in the mapping can be computed from the Lie derivatives according to the procedure spelled out below.

Theorem 5.16. Let $F(z_1, \ldots, z_n) = (F_1, \ldots, F_n)$ be a mapping from $\mathcal{N}$ to $\mathcal{N}$ which is covariant relative to a representation of a Lie group $\mathcal{G}$ with infinitesimal generators $J_1, \ldots, J_k$. Let the matrices of J_r on $\mathcal{N}$ be denoted by $(J_r)_{ij}$. The mapping F must satisfy

$$\sum_{j=1}^{n} (J_r)_{ij} F_j = J_r F_i \qquad r = 1, \ldots, k \tag{5.15}$$

where J_r acts on F_i as a derivation on the polynomial algebra $K[z_1, \ldots, z_n]$.

Proof. We have $T(g)F = FT(g)$ and so

$$\sum_{j=1}^{n} T_{ij}(g) F_j(z_1, \ldots, z_n) = F_i(T(g)z_1, \ldots, T(g)z_n).$$

Differentiating both sides along any one-parameter semigroup $g(t)$ through the identity $e \in \mathcal{G}$, we get

$$J_{ij} = \frac{d}{dt} T_{ij}(g(t))\Big|_{t=0} ,$$

$$\sum_{j=1}^{n} J_{ij} F_j = \frac{d}{dt} F_i(T(g(t))z_1, \ldots)\Big|_{t=0} = JF_i .$$

We get (5.15) by differentiating along all possible one-parameter groups.

Example 5.17. The action of SO(1) on the variables z and $\bar{z}$ is

$$T(\theta)z = e^{i\theta}z, \quad T(\theta)\bar{z} = e^{-i\theta}\bar{z}.$$

The Lie derivative $J = T'(0)$ is $Jz = iz$, $J\bar{z} = -i\bar{z}$. The polynomial $z^k\bar{z}^\ell$ is invariant if $Jz^k\bar{z}^\ell = 0$, hence if

$$\begin{aligned} Jz^k\bar{z}^\ell &= kz^{k-1}iz\bar{z}^\ell + z^k\ell\bar{z}^{\ell-1}(-i\bar{z}) \\ &= i(k-\ell)z^k\bar{z}^\ell = 0 ; \end{aligned}$$

so we must have $k = \ell$ and the invariants are $|z|^{2k}$, $k = 0, 1, \ldots$. A mapping $F_1(z,\bar{z})$ is covariant if

$$iF_1(z,\bar{z}) = JF_1$$

where J acts on the right as a derivation. The eigenfunctions of J are all of the form $z^k\bar{z}^\ell$ so F_1 must be a sum of terms of the form $z^k\bar{z}^\ell$ where $k-\ell = 1$:

$$\begin{aligned} F_1(z,\bar{z}) &= \sum_{\ell=0}^{\infty} a_\ell z^{\ell+1}\bar{z}^\ell \\ &= z\sum_{\ell=0}^{\infty} a_\ell |z|^{2\ell} \\ &= z\,a(|z|^2) \end{aligned}$$

as we found previously.

6. The Rotation Group SO(3).

The irreducible representations of the rotation group are denoted by D^{ℓ}, $\ell = 0, 1, 2, \ldots$ and are of dimension $2\ell + 1$. They are obtained when one considers the action of rotations on the spherical harmonics of degree ℓ:

$$Y_{\ell,m}(\theta,\varphi) = P_{\ell,m}(\theta)e^{im\varphi}$$

$$-\ell \leq m \leq \ell$$

where the $P_{\ell,m}$ are the associated Legendre functions. (See Wigner [7], p. 154 for their definition.)

Lemma 5.18. Each conjugacy class of SO(3) consists of all rotations through an angle of given magnitude.

Proof. The eigenvalues of every orthogonal matrix O are equal to one in absolute value; for if $O\underline{x} = \lambda\underline{x}$ then $\|\underline{x}\| = \|O\underline{x}\|^2 = \|\lambda\underline{x}\|^2 = |\lambda|\,\|\underline{x}\|^2$ and $|\lambda| = 1$. Furthermore, since O is real, its eigenvalues occur in complex conjugate pairs; so denote them by $\lambda, e^{i\varphi}$ and $e^{-i\varphi}$. Since $\det O = 1$, $\lambda e^{i\varphi}e^{-i\varphi} = \lambda = 1$.

Let $\underline{x}_1$ be the eigenvector of O with eigenvalue 1. Then $O\underline{x}_1 = \underline{x}_1$ and, since $O^+O = I$, $O^+\underline{x}_1 = \underline{x}_1$ as well. The orthogonal complement M of $[\underline{x}_1]$ is therefore also invariant under O.

When O is restricted to M it is still an orthogonal transformation with eigenvalues $e^{i\varphi}$ and $e^{-i\varphi}$, so its matrix must have the form

$$\begin{pmatrix} 1 & 0 & 0 \\ 0 & \cos\varphi & \sin\varphi \\ 0 & -\sin\varphi & \cos\varphi \end{pmatrix}$$

(Taking $\underline{x}_1 = \begin{pmatrix} 1 \\ 0 \\ 0 \end{pmatrix}$) In other words, a basis can be chosen in which O represents a rotation through an angle φ about a given axis. By making appropriate changes of bases every such orthogonal matrix O can be brought to the above form (by an orthogonal transformation) and so all elements $O \in SO(3)$ with eigenvalues $1, e^{i\varphi}, e^{-i\varphi}$ are conjugate.

The character of such an element O is $\chi = \mathrm{Tr}\, O = 1 + 2\cos\varphi$.

<u>Lemma 5.19.</u> The character of the ℓth representation is

$$\chi_\ell(\varphi) = \sum_{m=-\ell}^{\ell} e^{im\varphi} \tag{5.16}$$

<u>Proof.</u> Since the characters depend only on the conjugacy class it suffices to compute the trace of D^ℓ for a rotation through an angle φ about any fixed axis. In particular, the z-axis will do nicely. When we apply a rotation through an angle α about the z-axis to $Y_{\ell,m}(\theta,\varphi)$ we get

$$D_\ell(\alpha)Y_{\ell,m}(\theta,\varphi) = P_{\ell,m}(\theta)e^{im(\varphi-\alpha)}$$
$$= e^{-im\alpha}Y_{\ell,m}(\theta,\varphi) .$$

The matrix of $D_\ell(\alpha)$ is therefore

$$\begin{pmatrix} e^{-i\ell\alpha} & 0 & & \\ 0 & e^{-i(\ell-1)\alpha} & & 0 \\ \vdots & & \ddots & \\ & 0 & & e^{i\ell\alpha} \end{pmatrix}$$

and its trace is

$$\chi_\ell(\alpha) = \sum_{m=-\ell}^{\ell} e^{im\alpha}$$

Lemma 5.20. (Clebsch-Gordon series.) The tensor product $D^\mu \times D^\nu$ can be decomposed as follows:

$$D^\mu \otimes D^\nu = D^{\mu+\nu} \oplus \ldots \oplus D^{|\mu-\nu|} \tag{5.17}$$

Proof. It suffices to show that

$$\chi_\mu(\varphi)\chi_\nu(\varphi) = \chi_{\mu+\nu}(\varphi) + \ldots + \chi_{|\mu-\nu|}(\varphi) .$$

On the left we have

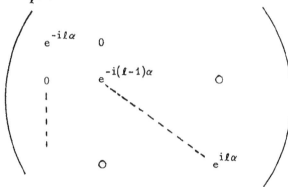

$$\sum_{m=-\mu}^{\mu} e^{im\varphi} \sum_{n=-\nu}^{\nu} e^{in\varphi} = \sum_{m=-\mu}^{\mu} \sum_{n=-\nu}^{\nu} e^{i(m+n)\varphi} ,$$

thus a sum of exponentials over the lattice of a rectangle of length μ and side ν. The problem is to rearrange the terms in this sum so that one is summeing $e^{ik\varphi}$ where k ranges from $\mu+\nu$ to $-\mu-\nu$. The details are given in Wigner, p. 186; we will give a closely related proof below using the Lie algebra of SO(3).

Take the operators L_1, L_2, and L_3 which satisfy the commutation relations (5.14) and form the new operators $J^{\pm} = \pm L_2 + iL_1$ and $J^3 = -iL_3$. These operators satisfy the commutation relations for $Sl(2)$

$$[J^+, J^-] = 2J^3 \qquad [J^3, J^{\pm}] = \pm J^{\pm} . \tag{5.17}$$

A representation of a Lie algebra on a vector space V is a mapping $J \to \rho(J)$ onto linear transformations on V such that

$$\rho(\alpha J_1 + \beta J_2) = \alpha\rho(J_1) + \beta\rho(J_2)$$

$$[\rho(J_1), \rho(J_2)] = \rho([J_1, J_2]) .$$

The representation is irreducible if there are no proper invariant subspaces of V; and the representation of the Lie algebra is irreducible if and only if the corresponding action on the Lie group on V is so. Consequently the irreducible representations of a Lie group can be constructed by finding the irreducible representations of its Lie algebra.

Theorem 5.21. Let V transform irreducibly under a representation of SO(3). Then $\dim V = 2\ell + 1$ and there exists a basis $\{f_m\}$ of the complexified space $V + iV$ such that

$$\begin{aligned} J_3 f_m &= m f_m \\ J_\pm f_m &= \beta_{\pm m} f_{m \pm 1} \end{aligned} \tag{5.18}$$

where $-\ell \leq m \leq \ell$ and $\beta_m = \sqrt{(\ell-m)(\ell+m+1)}$. In addition, the f_m can be normalized so that

$$\overline{f_m} = (-1)^m f_{-m} \; . \tag{5.19}$$

Proof. The construction of the basis $\{f_m\}$ can be carried out using only the commutation relations (5.17). See Miller, p. 234. Since I have not found a proof of (5.19) in the standard references, I will give one here. First, the operators L_1, L_2, L_3 are real operators, and so $\overline{J^3} = -J^3$, $\overline{J^{\mp}} = -J^-$, $\overline{J^-} = -J^+$. It follows that $\overline{J^3 f_m} = \overline{m f_m} = -J^3 \overline{f_m}$ and therefore that $J_3 \overline{f_m} = -m \overline{f_m}$. The vector $\overline{f_m}$ has weight $-m$; but since V is irreducible there is only one such vector, and that is f_{-m}. So $\overline{f_m} = c_m f_{-m}$. On the other hand, $J^+ f_m = \beta_m f_{m+1}$ so $\overline{J^+ f_m} = \beta_m \overline{f_{m+1}} = \beta_m c_{m+1} f_{-(m+1)}$ $= -J^- \overline{f_m} = -J^- c_m f_{-m} = -c_m J^- f_{-m} = -c_m \beta_m f_{-(m+1)}$. Consequently, $c_{m+1} = -c_m$ and we can take $c_m = (-1)^m c$. For $m = 0$ we have $\overline{f_0} = c f_0$. Choosing $c = 1$ we obtain that f_0 is real and $\overline{f_m} = (-1)^m f_{-m}$.

The reality condition (5.19) is important when we wish to restrict ourselves to real solutions of the bifurcation equations.

Now let us derive the Clebsch-Gordon series using Lie algebras. Suppose we have vector spaces V_μ and V_ν which transform as D^μ and D^ν respectively. Let them have bases $\{f_m\}$ and $\{g_n\}$ respectively, where $|m| \leq \mu$ and $|n| \leq \nu$. The tensor product space $V_\mu \otimes V_\nu$ is spanned by $\{f_m \otimes g_n\}$. The action of J_3 on these basis vectors is

$$J_3 f_m \otimes g_n = (J_3 f_m) \otimes g_n + f_m \otimes J_3 g_n = (m+n) f_m \otimes g_n ,$$

as $f_m \otimes g_n$ is an eigenvector of J_3 with weight $m+n$. Applying J_- to this vector we get

$$J_- f_m \otimes g_n = \beta_{-m} f_{m-1} \otimes g_n + \beta_{-n} f_m \otimes g_{n-1}$$

which is a vector of weight $m+n-1$. By successive application of J_- we get a string of vectors until finally the last vector is annihilated. The entire string of such vectors constitutes a subspace which is invariant and irreducible under the representation. We begin with the vector $f_\mu \otimes g_\nu$ and apply J_- to it successively. This generates an invariant subspace which transforms like $D^{\mu+\nu}$. This string containsone vector of weight $\mu+\nu-1$, namely

$$\beta^\mu_{-\mu} f_{\mu-1} \otimes g_\nu + \beta^\nu_{-\nu} f_\mu \otimes g_{\nu-1} .$$

But there are two vectors of weight $\mu+\nu-1$, so we may choose a second vector of weight $\mu+\nu-1$ which is orthogonal to this one and generate a second string. Proceeding in this way we generate invariant irreducible subspaces which transform like $D^{\mu+\nu-1}$, with $j = 0, 1, 2, \ldots$. When does it end? The process comes to an end when there are no more available vectors of a given weight. The dimension $n(j)$ of the subspace of vectors of weight $\mu+\nu-j$ can be determined from the figure below

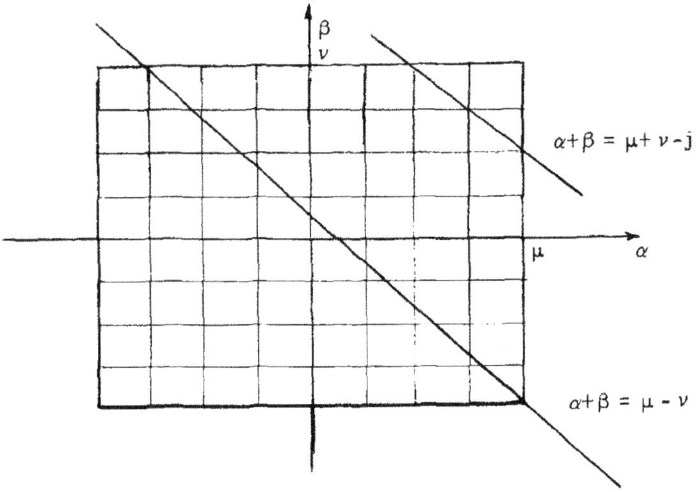

The dimension $n(j)$ is precisely equal to the number of lattice points in the rectange which lie on the line $\alpha+\beta = \mu+\nu-j$. Assume $\nu \leq \mu$. $n(0) = 1$, $n(1) = 2, \ldots, n(2\nu) = 2\nu+1$, but $n(2\nu+1) = 2\nu+1$, also. Therefore the process must terminate when $j = 2\nu$.

Just to make sure we have gotten everything let us check the dimensions. We have

$$\dim V_\mu \otimes V_\nu = (\dim V_\mu)(\dim V_\nu) = (2\mu+1)(2\nu+1).$$

On the other hand, a string whose highest weight is $\mu+\nu-j$ has length (dimension) $2(\mu+\nu-j)+1$; and summing these from $j = 0$ to $j = 2\nu$ we get

$$\sum_{j=0}^{2\nu} \{2[(\mu+\nu)-j]+j\}$$

$$= [2(\mu+\nu)+1][2\nu+1] - \sum_{j=1}^{2\nu} j$$

$$= (2(\mu+\nu)+1)(2\nu+1) - 2\nu(2\nu+1) = (2\mu+1)(2\nu+1).$$

The special case $\mu = \nu$ is important in our analysis of bifurcation equations. The highest weight vector is $f_\mu \otimes f_\mu$, which is symmetric. Since $J_\pm$ preserve the subspace of symmetric vectors, the entire string consists of symmetric vectors. When we go to the second string we choose a vector of weight $2\mu-1$ which is anti-symmetric, namely a multiple of $f_{\mu-1} \otimes f_\mu - f_\mu \otimes f_{\mu-1}$. Then the entire second string is anti-symmetric, and so on. Therefore

$$D^\mu \otimes D^\mu = D^{2\mu} \oplus \ldots \oplus D^0$$

where the first space is symmetric, the second anti-symmetric, and so forth. In particular, the subspace which transforms like D^μ is symmetric if μ is even and anti-symmetric if μ is odd. But this subspace of tensors corresponds to the corresponding covariant bilinear mapping. The mapping is symmetric if μ is even and anti-symmetric if μ is odd. Therefore there are no quadratic covariant terms for μ odd. A generalization of this result is given in Sattinger [5].

Construction of the Bifurcation Equations. Suppose now that the kernel $\mathcal{N}$ of L_0 transforms as D^ℓ and let us exhibit the algorithm for constructing the terms of various orders in the bifurcation equations. We identify $\mathcal{N}$ with linear polynomials in $z_{-\ell}, \ldots, z_\ell$ and the tensor product spaces over $\mathcal{N}$ with the algebra $K[z_{-\ell}, \ldots, z_\ell]$. The operators J_+, J_- and J_3 then act on $\mathcal{N}$ as

$$J_+ z_m = \beta_m z_{m+1}$$

$$J_- z_m = \beta_{-m} z_{m-1}$$

$$J_3 z_m = m z_m$$

and as derivations on the algebra K. Let the bifurcation equations take the form

$$F_m(z_{-\ell}, \ldots, z_\ell) = 0 \qquad m = -\ell, \ldots, \ell .$$

Since the representation is irreducible the linear terms are $F_m = \lambda z_m$. We get the quadratic terms as follows.

$$J_3(z_j z_k) = (J_3 z_j)z_k + z_j(J_3 z_k)$$

$$= (j+k)z_j z_k \ .$$

Therefore, since F_m must have weight m, that is $J_3 F_m = mF_m$, (see (5.15))

$$F_m = \sum_{j+k=m} a_{mjk} z_j z_k \ .$$

In particular,

$$F_\ell = a_\ell z_\ell z_0 + a_{\ell-1} z_{\ell-1} z_1 + \ldots$$

But $J_+ F_\ell = \beta_\ell F_\ell = 0$, applying this condition to F_ℓ we get a sequence of equations for $a_0, \ldots, a_\ell$ which uniquely determines these numbers up to a scalar multiple. For example, in the case $\ell = 2$ we have

$$J_+ F_2 = J_+(az_2 z_0 + bz_1^2)$$

$$= a\beta_0 z_2 z_1 + 2b\beta_1 z_1 z_2 \ .$$

Then a and b must satisfy $a\beta_0 + 2b\beta_1 = 0$, and this determines F_2 up to a constant scalar multiple. Once F_ℓ is obtained in this way, the other F_m's are obtained by applying J_- successively. For

$$J_- F_\ell = \beta_{-\ell} F_{\ell-1}$$

$$J_- F_{\ell-1} = \beta_{-\ell+1} F_{\ell-2}$$

and so forth. In this way the entire quadratic mapping is obtained. The coefficients obtained in this way are known as the Clebsch-Gordon coefficients. We write

$$F_m(z_{-\ell}, \ldots, z_\ell) = \sum_{m_1, m_2} c(\ell, m_1, \ell, m_2; \ell, m) z_{m_1} z_{m_2}$$

The Clebsch-Gordon coefficients are related to the Wigner 3_{-j} sybbols by

$$\begin{pmatrix} \ell & \ell & \ell \\ m_1 & m_2 & m_3 \end{pmatrix} = (-1)^{m_3} c(\ell, m_1, \ell, m_2; \ell, -m_3)$$

so we can also write

$$f_m(z_{-\ell}, \ldots, z_\ell) = \sum_{m_1+m_2+m=0} (-1)^m \begin{pmatrix} \ell & \ell & \ell \\ m_1 & m_2 & -m \end{pmatrix} z_{m_1} z_{m_2} .$$

The same procedure may be used to obtain covariant mappings of any degree. For example, to obtain covariant mappings of degree three, put

$$F_\ell = \sum_{i+j+k=\ell} a_{ijk} z_i z_j z_k .$$

Then $J_3 F_\ell = \ell F_\ell$ and the coefficients a_{ijk} are determined by setting $J_+ F_\ell = 0$. This leads to a system of equations for the a_{ijk}.

For example, when $\ell = 3$ we get

$$F_3 = az_3^2 z_{-3} + bz_3 z_2 z_{-2} + cz_3 z_1 z_{-1} + dz_3 z_0^2$$
$$+ ez_2 z_1 z_0 + fz_2^2 z_{-1} + gz_1^3$$

and the condition $J_+ F_3 = 0$ leads to five equation in seven unknowns:

$$\begin{bmatrix} \beta_{-3} & \beta_2 & 0 & 0 & 0 & 0 & 0 \\ 0 & \beta_{-2} & \beta_1 & 0 & 0 & 2\beta_2 & 0 \\ 0 & 0 & \beta_1 & 2\beta_1 & \beta_2 & 0 & 0 \\ 0 & 0 & 0 & 0 & \beta_1 & \beta_{-1} & 0 \\ 0 & 0 & 0 & 0 & \beta_0 & 0 & 3\beta_1 \end{bmatrix} \begin{bmatrix} a \\ b \\ c \\ d \\ e \\ f \\ g \end{bmatrix} = 0$$

There are two linearly independent solutions to this system, hence two linearly independent quadratic terms of degree 3:

$$F_m = z_m(z_0^2 - 2z_1 z_{-1} + 2z_2 z_{-2} - 2z_3 z_{-3})$$

which is obtained by taking $g = 0$ and $d = 1$. A second choice is $g \neq 0$, $d = 0$; the choice $g = \sqrt{7}$ leads to

$$G_3 = 9\sqrt{\frac{60}{7}}\, z_3^2 z_{-3} - 9\sqrt{\frac{60}{7}}\, z_3 z_2 z_{-2} + 3\sqrt{\frac{60}{7}}\, z_3 z_1 z_{-1}$$
$$-3\sqrt{10}\, z_2 z_1 z_0 + \frac{30}{\sqrt{7}}\, z_2^2 z_{-1} + \sqrt{7}\, z_1^3 ,$$

with the lower weight polynomial obtfined by successive applications of the lowering operator J_-.

Variational Structure. When ℓ is even the quadratic terms possess a gradient structure, as follows. The second order polynomial

$$q(z_{-\ell},\ldots,z_\ell) = \sum_{m=-\ell}^{\ell} (-1)^m z_m z_{-m}$$

is an invariant under the action of SO(3). In fact, one can easily see that J_+, J_-, and J_3 all annihilate q; or, note that q is really the square of the Euclidean norm on the original real vector space $\mathcal{H}$, and so is invariant since the group operations are all orthogonal transformations on $\mathcal{H}$. When q is restricted to the real subspace given by $\bar{z}_m = (-1)^m z_{-m}$ it has the value

$$|z|^2 = q(z_{-\ell},\ldots,z_\ell) = \sum_{m=-\ell}^{\ell} |z_m|^2 .$$

The homogeneous third-degree polynomial

$$p(z_{-\ell},\ldots,z_\ell) = \frac{1}{3}\sum_{-\ell}^{\ell} F_m \bar{z}_m = \frac{1}{3}\sum_{-\ell}^{\ell} (-1)^m F_m z_{-m}$$

is also an invariant; again one can check that p is annihilated by the three infinitesimal generators J_+, J_-, and J_3. Furthermore, from the representation for F_m,

$$p = \frac{1}{3} \sum_{m_1+m_2+m_3=0} \begin{pmatrix} \ell & \ell & \ell \\ m_1 & m_2 & m_3 \end{pmatrix} z_{m_1} z_{m_2} z_{m_3} .$$

For even ℓ the 3_{-j} symbols are completely symmetric in m_1, m_2, and m_3 (Hammermesh, p. 159) and therefore

$$\frac{\partial p}{\partial z_m} = F_m(z_{-\ell}, \ldots, z_\ell).$$

The reduced bifurcation equations take the form

$$\tau z_m + \frac{\partial p}{\partial z_m} = 0 ,$$

and these are the Euler-Lagrange equations for the variational problem

$$\operatorname*{Min}_{q=1} p$$

For further discussion of variational methods, see Sattinger [5].

References

1. Hammermesh, M. Group Theory and its Applications to Physical Problems, Addison-Wesley, Reading, Mass. 1962.

2. Jaric, M.V., and Birman, J. L., "New algorithms for the Molien functions," Jour. Math. Phys. 18 (1977), 1456-1458.

3. Miller, W., Symmetry Groups and their Applications, Academic Press, New York, 1972.

4. Sattinger, D.H., "Group theory and branch points of nonlinear functional equations," SIAM Jour. Math. Anal. 8 (1977), 179-201.

5. ______, "Bifurcation from rotationally invariant states," Jour. Math. Phys. 19 (1978), 1720-1732.

6. Springer, T.A., Invariant Theory, Springer Lecture Notes in Mathematics, no. 585 (1977).

7. Wigner, E., Group Theory and Atomic Spectra, Academic Press, New York, 1959.

VI

APPLICATIONS

6.1. Breaking of Euclidean Invariance

In this chapter we discuss in some detail the application of group-theoretic methods to some well-known problems in applied mathematics. We begin by discussing the Bénard problem, which was mentioned in the beginning of the Introduction; but we shall discuss it here from the more general point of view of the breaking of Euclidean invariance in the plane.

Suppose the equations (1.1) are covariant with respect to a representation of the group of rigid motions $\mathcal{E}(2)$ in the plane. If u is a scalar valued function, the appropriate representation is $(T_g u)(\underline{x}) = u(g^{-1}\underline{x})$, where $g\underline{x} = O\underline{x} + a$, O being an orthogonal matrix. In the case of the convection problems, $u = (u_1, u_2, u_3, \theta, p)$ where u_j are the components of the velocity field, θ is the temperature, and p is the hydrodynamic pressure. In this case the appropriate representation is given by

$$(T_g u)(\underline{x}) = \left(\begin{array}{ccc|cc} & & & 0 & \\ & O & & 0 & \\ & & & 0 & \\ \hline & & & 1 & \\ & O & & & 1 \end{array}\right)\begin{pmatrix} u_1 \\ u_2 \\ u_3 \\ \theta \\ p \end{pmatrix}(g^{-1}\underline{x})$$

The reader may check that the Boussinesq equations are covariant with respect to this representation. In general, we assume that u is a function on the plane taking its values in a Banach space $\mathcal{E}$, and that the representation of $\mathcal{E}(2)$ takes the form $(T_g u)(\underline{x}) = (S_g u)(g^{-1}\underline{x})$ where S_g is a representation of $\mathcal{E}(2)$ on $\mathcal{E}$.

Assume $G(\lambda, 0) \equiv 0$ and let $L(\lambda) = G_u(\lambda, 0)$ be the linearization of the equations about the zero solution. Now $L(\lambda)$ is covariant with respect to the entire group of rigid motions, and so in particular it commutes with translations. Therefore its invariant subspaces are of the form $\psi_\omega = ve^{i<\underline{\omega}, \underline{x}>}$, where $\underline{\omega}, \underline{x} \in \mathbb{R}^2$ and $v \in \mathcal{E}$. The stability of the zero solution is formally determined by the eigenvalue problem $L(\lambda)ve^{i<\underline{\omega}, \underline{x}>} = \sigma(\lambda, \underline{\omega})ve^{i<\underline{\omega}, \underline{x}>}$. Since the problem is rotationally invariant, σ depends only on $w = |\underline{\omega}|$. The criticality condition is given by a relation of the form $\sigma(\lambda, \omega) = 0$ where σ is the principal eigenvalue. The curve $\sigma(\lambda, \omega) = 0$ is called the <u>neutral stability curve</u> and determines the relationship between the critical parameter value λ_c and the critical wave number ω_c. A typical neutral stability curve, such as that which occurs in the Bénard problem, is pictured in Figure 6.1. The critical

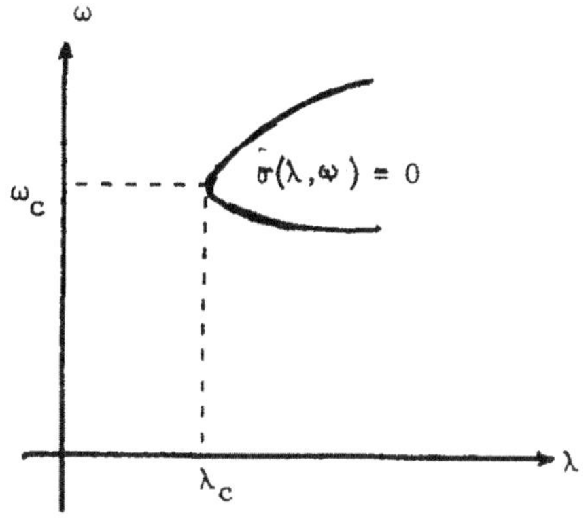

Figure 6.1

parameter value λ_c is that for which infinitesimal disturbances of some wave number first go unstable. That critical wave number ω_c then determines the wavelength of the bifurcating disturbances.

Since $L(\lambda_c)$ is covariant with respect to rotations, its kernel consists of the entire circle of wave functions

$$\{(S_r v)\, e^{i<\underline{\omega},\underline{x}>}\}_{r \in O(2)} \tag{6.1}$$

This vector space is infinite-dimensional but the multiplicity reduces to a finite one if we restrict ourselves to a subclass of doubly periodic functions. Let Λ be a lattice of vectors in the plane: that is, if $\underline{\omega}_1$ and $\underline{\omega}_2$ belong to Λ then so do $n\underline{\omega}_1 + m\underline{\omega}_2$ for all $m, n \in \mathbb{Z}$. Let $H(\Lambda)$ denote the subgroup of translations through vectors in Λ

$$H(\Lambda) = \{T\underline{\omega} \mid \underline{\omega} \in \Lambda\} .$$

The set of all functions ψ which are invariant under $H(\Lambda)$: $T_\omega \psi = \psi$ for all $\omega \in \Lambda$ is called the class of Λ-periodic functions and is denoted by $\mathcal{J}(\Lambda)$. $\mathcal{J}(\Lambda)$ is invariant under the mapping $G(\lambda, u)$

since G commutes with translations, and so we may restrict ourselves to the study of bifurcating doubly periodic solutions. On the subspace $\mathcal{J}(\Lambda)$ the kernel $\mathcal{N}(\Lambda)$ of $L(\lambda_c)$ is finite-dimensional, being equal to the number of wave functions in (6.1) which are Λ-periodic. By such a procedure we obtain bifurcating solutions which break Euclidean invariance; it is an open question, however, whether this is the only way Euclidean invariance can be broken.

A wave function $e^{i<\underline{\omega}', \underline{x}>}$ is Λ-periodic if and only if $e^{i<\underline{\omega}', \underline{x}+\underline{\omega}>} = e^{i<\underline{\omega}', \underline{x}>} e^{i<\underline{\omega}', \underline{\omega}>} = e^{i<\underline{\omega}', \underline{x}>}$; hence the condition that $e^{i<\underline{\omega}', \underline{x}>}$ be Λ-periodic is that

$$e^{i<\underline{\omega}', \underline{\omega}>} = 1, \quad \text{or} \quad <\underline{\omega}', \underline{\omega}> = 2\pi(\text{integer})$$

for all $\underline{\omega} \in \Lambda$. This condition describes the dual lattice Λ':

$$\Lambda' = \{ \underline{\omega}' \mid <\underline{\omega}', \underline{\omega}> = 2\pi(\text{integer}) \text{ for all } \underline{\omega} \in \Lambda \}.$$

Therefore $\mathcal{N}(\Lambda)$ consists of all wave functions in (6.1) for which $r\underline{\omega}$ lies in the dual lattice Λ'. Of course, $\mathcal{N}(\Lambda)$ may be empty for some lattices Λ_j and in that case there are no bifurcating solutions with that periodicity. If $e^{i<\underline{\omega}', \underline{\omega}>} \in \mathcal{J}(\Lambda)$ then $T_r e^{i<\underline{\omega}', \underline{x}>} = e^{i<\underline{\omega}', r^{-1}\underline{x}>} = e^{i<r\underline{\omega}', \underline{x}>} \in \mathcal{J}(\Lambda)$ if and only if the rotation r leaves Λ' invariant. The largest subgroup of O(2) which leaves Λ' invariant is called the <u>holohedry</u> $\mathcal{H}(\Lambda')$ of the lattice Λ'. The kernel $\mathcal{N}(\Lambda)$ is then given by

$$\mathcal{N}(\Lambda) = \{ (S_r v) e^{i \langle r\underline{\omega}', \underline{x} \rangle} \}_{r \in \mathcal{D}(\Lambda')}$$

This set is finite, though not necessarily equal to the number of elements in $\mathcal{D}(\Lambda')$. For example, in the case of the hexagonal lattice $\mathcal{D}(\Lambda') = D_6$ has order 12 while there are 6 wave functions $e^{i\langle \underline{\omega}_j, \underline{x} \rangle}$. On the other hand, one also has to take into account how the $S_r v$ transform.

For in the vector valued case $\mathcal{N}(\Lambda)$ is the tensor product of two spaces. One is the vector space spanned by the $e^{i\langle \underline{\omega}_j, \underline{x} \rangle}$ and the other is spanned by the vectors $(S_r v)$. The representation on $\mathcal{N}(\Lambda)$ is the tensor product of S_r and the action of $\mathcal{D}(\Lambda')$ on the $e^{i\langle \underline{\omega}_j, \underline{x} \rangle}$. In the case of the Bénard problem with free-free boundary conditions, it can be shown (Sattinger [6]) that the kernel $\mathcal{N}(\Lambda)$ transforms as do the scalar fields $\psi_j(\underline{x}) = e^{i\langle \underline{\omega}_j, \underline{x} \rangle}$. But there is no a priori reason for this always to be the case. In some physical problems the kernel $\mathcal{N}(\Lambda)$ may transform in a more complicated fashion. The situation is similar to that of spin-angular momentum coupling in quantum mechanics.

We denote by $\mathcal{E}(\Lambda)$ the subgroup of rigid motions which leaves $\mathcal{F}(\Lambda)$ invariant. It contains all translations and the holohedry $\mathcal{D}(\Lambda')$.

Before discussing the structure of the bifurcation equations we determine the action of $\xi(\Lambda)$ on the kernel $\eta(\Lambda)$. We restrict ourselves to lattices generated by two vectors $\underline{\omega}_1$ and $\underline{\omega}_2$ whose length is ω_c, the critical wave length. These are pictured in Figure 6.2.

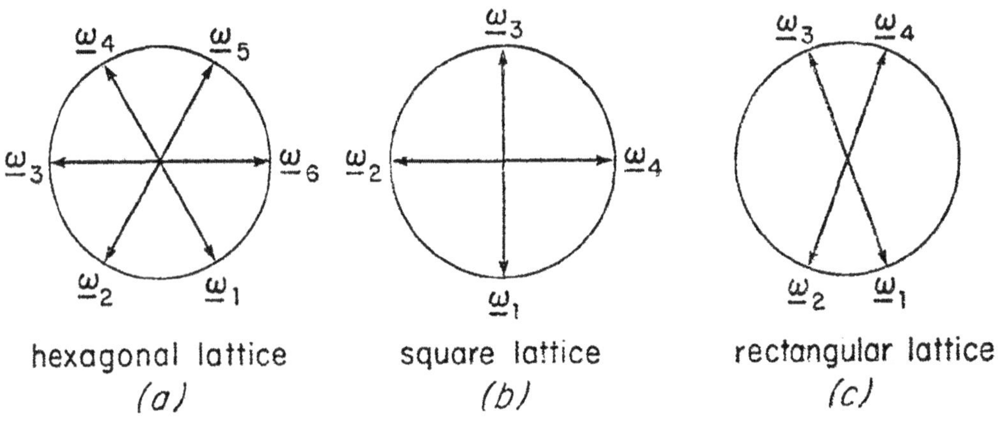

Figure 6.2

We consider the scalar case only. Let the wave functions in $\eta(\Lambda)$ be denoted by $\psi_j(\underline{x}) = e^{i<\underline{\omega}_j, \underline{x}>}$. The vertices of the wave vectors $\underline{\omega}_1, \underline{\omega}_2, \ldots$ joined together make up a regular polygon, which we call the fundamental polygon. In the more general, vector valued, case, the kernel $\eta(\Lambda)$ may transform in a more complicated way.

<u>Theorem 6.1.</u> <u>Let T_σ be the representation of the group of rigid motions restricted to the subgroup</u> $\mathcal{E}(\Lambda)$. <u>Then the action of T_σ on</u> $\mathcal{N}(\Lambda)$ <u>is as follows</u>

(i) If $\sigma = a$ is a translation

$$T_a \psi_j = e^{i<\underline{\omega}_j, \underline{a}>} \psi_j$$

(ii) <u>If $\sigma = r$ is a rotation/reflection in</u> $\mathcal{B}(\Lambda)$ <u>then T_r permutes the ψ_j in the same fashion that r permutes the vertices of the fundamental polygon. Specifically, let $r(j)$ be the permutation of the vertices $\underline{\omega}_j$ of the fundamental polygon. Then</u>

$$T_r \psi_j = \psi_{r(j)} .$$

(iii) $T_r \psi_j = \overline{\psi_j}$ <u>whenever</u> $r \underline{\omega}_j = -\underline{\omega}_j$.

The proof of Theorem 6.1 is immediate and therefore left to the reader.

Consider the hexagon in Figure 6.2. Its symmetry group is generated by the permutations

$$\alpha = (1\,2\,3\,4\,5\,6) \quad \text{and} \quad \beta = (2\,6)(3\,5).$$

The corresponding operations in $\mathcal{B}(\Lambda)$ are, respectively, a rotation through 60° and a reflection through the axis joining $\underline{\omega}_1$ and $\underline{\omega}_4$. The resulting transformations are

$$T(\alpha)\psi_1 = \psi_2 \quad , \quad T(\alpha)\psi_2 = \psi_3, \ldots$$

$$T(\beta)\psi_1 = \psi_1 \quad , \quad T(\beta)\psi_2 = \psi_6, \ldots$$

If w is a vector in $\mathcal{N}(\Lambda)$ we write

$$w = \sum_{j=1}^{6} z_j \psi_j .$$

If w is to be real we must require that $z_j = \overline{z}_k$ whenever $\psi_j = \overline{\psi_k}$, hence whenever $\underline{\omega}_j = -\underline{\omega}_k$. Thus

$$z_1 = \overline{z}_4 , \quad z_2 = \overline{z}_5 , \quad z_3 = \overline{z}_6 .$$

6.2. Construction of the Bifurcation Equations

To construct the covariant bifurcation equations for a given lattice Λ we first identify $\mathcal{N}(\Lambda)$ with linear polynomials in $z_1, \ldots, z_n$, where $n = \dim \mathcal{N}(\Lambda)$. In the case of the hexagonal lattice the group action on the variables is

$$\begin{aligned}
\alpha(z_1, \ldots, z_6) &= (z_2, z_3, \ldots, z_6, z_1) \\
\beta(z_1, \ldots, z_6) &= (z_1, z_6, z_5, z_4, z_3, z_2) \\
T_a(z_1, \ldots, z_6) &= (e^{i<\underline{\omega}_1, a>} z_1, \ldots, e^{i<\underline{\omega}_6, a>} z_6) \\
J(z_1, \ldots, z_6) &= (\overline{z}_1, \ldots, \overline{z}_6) .
\end{aligned}$$

The equations are covariant with respect to the operation J of complex conjugation since the original system G is real -- that is, G commutes with complex conjugation.

The mapping F is given by $F = (F_1, \ldots, F_6)$. From $\alpha F = F\alpha$ we obtain

$$F_2(z_1, \ldots, z_6) = F_1(z_1, \ldots, z_6) \qquad (6.2)$$

and so forth. Thus knowing the component F_1 we may obtain the others by cyclic permutation of the variables $z_1, \ldots, z_6$. From $\beta F = F\beta$ we get

$$F_1(z_1, \ldots, z_6) = F_1(z_1, z_6, z_5, z_4, z_3, z_2, z_1) \qquad (6.3)$$

Similarly, $JF = FJ$ implies

$$\overline{F_1(z_1, \ldots, z_6)} = F_1(\overline{z}_1, \ldots, \overline{z}_6)$$

Finally, from the translational invariance of F we get

$$e^{i<\underline{\omega}_1, \underline{a}>} F_1(z_1, \ldots, z_6) = F_1(e^{i<\underline{\omega}_1, \underline{a}>} z_1, \ldots, e^{i<\underline{\omega}_6, \underline{a}>} z_6) \qquad (6.5)$$

We may compute the general covariant mappings F from (6.2), (6.3), (6.4), (6.5). We decompose F_1 into linear, quadratic, cubic terms, etc. First consider the linear terms. If F_1 contains a term z_j then from (6.5)

$$e^{i<\underline{\omega}_1, \underline{a}>}(z_j) = (e^{i<\underline{\omega}_j, \underline{a}>} z_j) ;$$

but this equation can only hold for arbitrary $\underline{a}$ if $\underline{\omega}_j = \underline{\omega}_1$, so the only covariant linear term is

$$F_1 = az_1 .$$

Then $F_2 = az_2$, $F_3 = az_3, \ldots$. Consequently ,

Theorem 6.2. The kernel $\mathcal{N}(\Lambda)$ transforms irreducibly under the representation $T\sigma$.

Proof. For the case of the hexagonal lattice we have seen that the only linear covariant mapping is a scalar multiple of the identity. Therefore, by Schur's theorem (Theorem 5.2), $\mathcal{N}(\Lambda)$ is irreducible. The same proof holds for the other lattices as well.

In the case of quadratic terms the expression $z_j z_k$ is covariant only if

$$e^{i\langle \underline{\omega}_1, \underline{a}\rangle}(z_j z_k) = (e^{i\langle \underline{\omega}_j, \underline{a}\rangle} z_j)(e^{i\langle \underline{\omega}_k, \underline{a}\rangle} z_k)$$

for all $\underline{a}, z_j$, and z_k. This can be so if and only if

$$\underline{\omega}_1 = \underline{\omega}_j + \underline{\omega}_k ,$$

and this equation is satisfied only by taking $\underline{\omega}_j = \underline{\omega}_2$ and $\underline{\omega}_k = \underline{\omega}_6$. Thus the only covariant quadratic mapping is

$$F_1 = bz_2 z_6 , \quad F_2 = bz_3 z_1, \ldots$$

with b real. (The reality of b follows from (6.4)).

In the same way, the cubic term $z_j z_k z_\ell$ is covariant if and only if

$$\underline{\omega}_1 = \underline{\omega}_j + \underline{\omega}_k + \underline{\omega}_\ell .$$

This condition is met only if one vector on the right is $\underline{\omega}_1$ and the other two sum to zero. There are three possibilities, and the covariant terms are

terms are

$$z_1^2 z_4 , \quad z_1 z_2 z_5 , \text{ and } \quad z_1 z_3 z_6 .$$

However, from the symmetry $\beta F = F\beta$ it follows that F_1 is symmetric in the variables z_2, z_6 and z_3, z_5. Therefore F_1 takes the form

$$F_1 = cz_1^2 z_4 + d(z_3 z_6 + z_2 z_5).$$

The covariant bifurcation equations for the other lattices Λ_2 and Λ_4 are obtained in similar fashion. The derivation of the mapping F from F_1 in the case of the rhombic lattice Λ_2 differs slightly. The elements of D_2 are $\alpha = (1\,3)(2\,4)$, $\beta = (1\,2)\,1\,3\,4)$, and $\gamma = (2\,3)(1\,4)$, and the components F_2, F_3, and F_4 are obtained from F_1 by applying the symmetries $\alpha F = F\alpha$, $\beta F = F\beta$, and $\gamma F = F\gamma$. Below is a table for the generators F_1 for the covariant mappings of degree k, where $k = 1, 2, 3$.

k	Λ_2	Λ_4	Λ_6
1	az_1	az_1	az_1
2	0	0	$bz_2 z_6$
3	$cz_1^2 z_3 + dz_1 z_2 z_4$	$cz_1^2 z_3 + dz_1 z_2 z_4$	$cz_1^2 z_4 + dz_1(z_2 z_5 + z_3 z_6)$

Table 6.1

The coefficients c and d in the lattices Λ_2, Λ_4, and Λ_6 are not the same, but depend on the lattice. The nature of this dependency will be discussed below. The bifurcation of doubly periodic solutions in each of the lattices can be analyzed by straightforward algebraic methods, and the formal (linearized) stability of the solutions can be determined. We point out here an important simplification of the bifurcation equations that can be obtained by the introduction of "action-angle" variables. For Λ_6 these variables are

$$z_j = x_j e^{i\theta_j}, \quad z_{j+3} = x_j e^{-i\theta_j}, \quad j = 1, 2, 3$$

where $x_j \geq 0$ and $0 \leq \theta_j < 2\pi$. The group action in these coordinates is given by

$$\alpha(x_1, x_2, x_3; \theta_1, \theta_2, \theta_3) = (x_2, x_3, x_1; \theta_2, \theta_3, -\theta_1)$$

$$\beta(x_1, x_2, x_3; \theta_1, \theta_2, \theta_3) = (x_1, x_3, x_2; \theta_1, -\theta_3, -\theta_2)$$

$$T_a(x_1, x_2, x_3; \theta_1, \theta_2, \theta_3) = (x_1, x_2, x_3; \theta_1 + \langle \underline{\omega}_1, \underline{a} \rangle, \theta_2 + \langle \underline{\omega}_2, \underline{a} \rangle, \theta_3 + \langle \underline{\omega}_3, \underline{a} \rangle)$$

When one introduces these variables the bifurcation equations simplify; in fact, their order is cut in half.

For example, suppose in the case Λ_6 we take for the reduced bifurcation equations the linear, quadratic and cubic terms, and substitute in the action-angle variables. We get

$$F_1 = ax_1e^{i\theta_1} + bx_2x_3e^{i(\theta_2-\theta_3)} + cx_1^3e^{i\theta_1} + dx_1(x_2^2+x_3^2)e^{i\theta_1}$$

$$F_2 = ax_2e^{i\theta_2} + bx_3x_1e^{i(\theta_1+\theta_3)} + cx_2^3e^{i\theta_2} + dx_2(x_3^2+x_1^2)e^{i\theta_2}$$

$$\vdots$$

Dividing the above equations by $e^{i\theta_1}$, $e^{i\theta_2}$, etc. we get

$$F_1 = ax_1 + bx_2x_3e^{i(\theta_2-\theta_3-\theta_1)} + cx_1^3 + dx_1(x_2^2+x_3^2)$$

$$F_2 = ax_2 + bx_3x_1e^{i(\theta_1+\theta_3-\theta_2)} + cx_2^3 + dx_2(x_1^2+x_2^2)$$

$$\vdots$$

Given any solution $(x_1, x_2, x_3; \theta_1, \theta_2, \theta_3)$ we may translate it through some vector $\underline{a}$ to make $\theta_1 = \theta_3 = 0$. Then θ_2 is replaced by $\theta_2 - \theta_1 - \theta_3$ since $\underline{\omega}_2 = \underline{\omega}_1 + \underline{\omega}_3$. Furthermore, the last three equations are simply complex conjugates of the first three. Letting $\gamma = \theta_2 - \theta_1 - \theta_3$ we get the simplified equations

$$F_1 = ax_1 + be^{i\gamma}x_2x_3 + cx_1^3 + dx_1(x_2^2+x_3^2) = 0\ ,$$

$$F_2 = ax_2 + be^{i\gamma}x_3x_1 + cx_2^3 + dx_2(x_1^2+x_3^2) = 0\ ,$$

$$F_3 = ax_3 + be^{i\gamma}x_1x_2 + cx_3^3 + dx_3(x_1^2+x_2^2) = 0\ .$$

γ is an arbitrary parameter, but since all other quantities are real, γ **must** be either 0 or π; and it doesn't matter which choice we take, since this simply determines the direction of parametrization of the transcritical bifurcation branch when $b \neq 0$. When we take $\gamma = 0$, the resulting equations are covariant with respect to the symmetry group S_3. Thus we have factored out the continuous group and are left with a problem covariant with respect to a discrete group. It is an interesting question whether such a procedure can be carried out for continuous groups in general.

A similar reduction of the equations can be carried out for the other lattices as well. The reduced bifurcation equations for the bifurcation of Λ periodic disturbances of a Euclidean covariant system are given below:

1. Λ = rectangular or square lattice

$$\begin{aligned} x_1(\lambda + cx_1^2 + dx_2^2) &= 0 \\ x_2(\lambda + cx_2^2 + dx_1^2) &= 0 \end{aligned} \tag{6.6}$$

2. Λ = hexagonal lattice $(k = 2)$

$$\begin{aligned} \lambda x_1 - x_2x_3 &= 0 \\ \lambda x_2 - x_3x_1 &= 0 \\ \lambda x_3 - x_1x_2 &= 0 \end{aligned} \tag{6.7}$$

3. Λ = hexagonal lattice $(k = 3)$

$$\begin{aligned} x_1(\lambda + cx_1^2 + d(x_2^2 + x_3^2)) &= 0 \\ x_2(\lambda + cx_2^2 + d(x_3^2 + x_1^2)) &= 0 \\ x_3(\lambda + cx_3^2 + d(x_1^2 + x_2^2)) &= 0 \end{aligned} \tag{6.8}$$

A complete analysis of these equations is a matter of straightforward algebra.

Equations (6.6) possess two types of solutions. If $x_2 = 0$ then they reduce to $\lambda + cx_1^2 = 0$, which has the unique solution $x_1 = \sqrt{-\lambda/c}$ (assuming $c \neq 0$). Similarly, one could take $x_1 = 0$ and $x_2 = \sqrt{-\lambda/c}$, but this is obtained by a symmetry operation from the first. The second type of solution is $x_1 = x_2 = \sqrt{-\lambda/(c+d)}$, provided $c+d \neq 0$. If $c = d$ then the equations reduce to $\lambda + c(x_1^2 + x_2^2) = 0$, which has the one-parameter family of solutions $x_1 = \sqrt{-\lambda/c}\cos\alpha$, $x_2 = \sqrt{-\lambda/c}\sin\alpha$; but this is a singular set.

Equations (6.7) have only one non-trivial solution, $x_1 = x_2 = x_3 = |\lambda|$. For if any x_i vanishes, they all do. If none of them vanish, multiply the three equations together to get $\lambda^3(x_1x_2x_3) = (x_1x_2x_3)^2$. From this it follows that $x_1x_2x_3 = \lambda^3$; and from the first equation we get $\lambda x_1^2 = x_1x_2x_3 = \lambda^3$, hence $x_1^2 = \lambda^2$, $x_1 = |\lambda|$, etc.

Equations (6.8) possess three types of solutions.

(a) $x_2 = x_3 = 0,\ x_1 = \sqrt{-\lambda/c}$; (b) $x_3 = 0,\ x_1 = x_2 = \sqrt{-\lambda/(c+d)}$;

(c) $x_1 = x_2 = x_3 = \sqrt{-\lambda/(c+2d)}$. If $c = d$ there are also singular solutions of the type $x_3 = 0,\ x_1 = \sqrt{-\lambda/c}\cos\alpha,\ x_2 = \sqrt{-\lambda/c}\sin\alpha$, etc.

To determine the stability of the bifurcating solutions we must compute the eigenvalues of the Jacobian of the reduced bifurcation equations. We have cut the equations in half in each of the above equations, but in cases (6. 6) and (6.8) it suffices to compute the Jacobians of those equations directly. For the details, see Sattinger [6]. The stability results are summarized in

Theorem 6.3. *The following are necessary conditions for the stability of the bifurcating solutions.*

(6.6) stable rolls : $d < c < 0$

stable squares/rectangles : $c + d < 0$, $c - d < 0$

(6.8) stable rolls : $d < c < 0$

stable hexagons : $c < d$, $c + 2d < 0$

(6.7) *In this case the hexagons bifurcate transcritically with both sides unstable; the subcritical branch has one unstable mode.*

6.3 Mechanisms for Pattern Selection

The stability results above refer to the stability of the given solutions with respect to disturbances within the same lattice class. Since there is a two-parameter translation group acting on the solutions there are in general two (one for the rolls) neutral modes present; so the stability is at best an orbital stability. The situation can be compared to the stability of time periodic solutions of an autonomous system of ordinary differential equations ($\dot{x} = f(x)$); the invariance of the equations under the translations $t \to t + \gamma$ means that one of the Floquet exponents is always zero. In the case (6.7) the hexagonal solutions bifurcate transcritically. The stability of this branch can be determined by an analysis of the Jacobian of the reduced bifurcation equations which are generated by cyclic permutation in the equation

$$\lambda z_1 - z_2 z_6 = 0 .$$

The unique orbit of solutions is generated by the solution $z_1 = \dots = z_6 = \lambda$. The Jacobian of the system at this solution is

$$\lambda \begin{bmatrix} 1 & -1 & 0 & 0 & 0 & -1 \\ -1 & 1 & -1 & 0 & 0 & 0 \\ 0 & -1 & 1 & -1 & 0 & 0 \\ 0 & 0 & -1 & 1 & -1 & 0 \\ 0 & 0 & 0 & -1 & 1 & -1 \\ -1 & 0 & 0 & 0 & -1 & 1 \end{bmatrix} \qquad (6.9)$$

and the eigenvalues of this matrix are $-\lambda, 3\lambda, 2\lambda, 2\lambda, 0, 0$. For supercritical bifurcation $(\lambda > 0)$ there are three unstable modes, two neutral modes and one stable mode; while for $\lambda < 0$ there is only one unstable mode. From Theorem 4.3 therefore, the bifurcating hexagonal solutions have one unstable mode subcritically. (Remark: Theorem 4.3 guarantees information only at lowest order, and since two of the eigenvalues of the Jacobian are zero, nothing can be concluded about the perturbed behavior of these eigenvalues by Theorem 4.3 alone. However, the presence of these two zero eigenvalues is due to the two-parameter translation group acting, and so they persist under the perturbation.)

The invariant subspaces of (6.9) may be found by decomposing the kernel $\mathcal{N}$ into irreducible invariant subspaces under the action of D_6 on the wave vectors $\{\psi_1, \ldots, \psi_6\}$. The matrix (6.9) commutes with the representation of D_6 since it is the Jacobian of the bifurcation equations at a solution which is invariant under the entire symmetry group of the hexagon ($T_g F(v) = F(T_g v)$; therefore $T_g F'(v) = F'(T_g v) T_g$, but if v is invariant under the entire group then $T_g F'(v) = F'(v) T_g$.) Therefore the invariant subspaces of (6.9) are the same as those of the representation of D_6 on $\mathcal{N}$. This representation and its irreducible invariant subspaces are given in [6].

Since the hexagonal solutions have one subcritical unstabl mode there is the possibility that the subcritical branch bends back and regains stability, as in the diagram below.

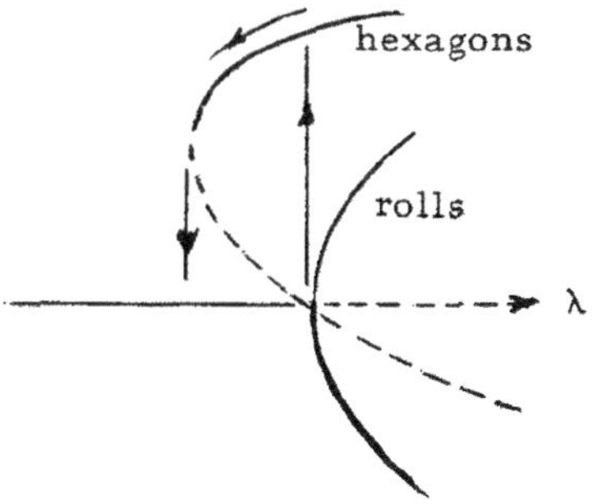

Figure 6.3

Should this be the case there is the possibility of discontinuous transitions at the critical point and also of hysteresis effects, as indicated by the arrows in the diagram.

In order to compare the stability of the bifurcating solutions in the other cases, one must be able to determine the coefficients c and d as functions of the lattice. We turn to that problem now. We are going to prove

Theorem 6.4. *There exists a function* $q(\theta)$, *called the lattice function, such that*

1) $q(\theta) = A_0 + A_2 \cos 2\theta + A_4 \cos 4\theta + \ldots$

2) $c = 3q(0)$, $d = 6q(\alpha)$

where α *is the acute angle between two adjacent basic vectors of the lattice, and the coefficients* $A_0, A_2, \ldots$ *depend only on the basic physical parameters of the problem.*

We prove Theorem 6.4 in full generality, for the case where u is not a scalar valued function on $\mathbb{R}^2$ but takes its values in a Banach space. Since we want to compare the coefficients c and d in the bifurcation equations for different lattices our first task is to construct a universal mapping F which, when restrictricted to any given lattice, yields the bifurcation equations for that lattice. In the following we deal with formal algebraic quantities. Let Σ denote the vector space of trigonometric polynomials in $\mathbb{R}^2$ with coefficients in $\mathcal{E}_1$. Thus, $\varphi \in \Sigma$ if

$$\varphi = \sum_{j=1}^{N} A_j e^{i\langle \underline{\omega}_j, \underline{x}\rangle}$$

where the vectors $\underline{\omega}_j$ comprise an arbitrary set of vectors in $\mathbb{R}^2$ and the elements A_j belong to $\mathcal{E}_1$. Now let $\mathcal{N} = \ker G_u(0,0)$ and denote by P the projection of Σ onto $\Sigma' = \Sigma \cap \mathcal{N}$. The operation P on φ consists imply in casting out all terms in the sum which do not lie in $\mathcal{N}$. We set $Q = I - P$.

Recall that the kernel has the form

$$\mathcal{N} = \{S_\theta \underline{v} e^{i<\underline{\omega}_\theta, \underline{x}>}\}_{0 \leq \theta < 2\pi}$$

where $\underline{v} e^{i<\underline{\omega}, \underline{x}>}$ is one element of $\ker G_u(0,0)$ and $\underline{\omega}_\theta$ is the vector $\underline{\omega}$ rotated through an angle θ. Let H be a (formally) analytic operator defined on Σ. We say that

$$H(\varphi) \equiv 0 \pmod{k}$$

if $H(z\varphi)$ has a kth order zero at $z = 0$; i.e., if $H(z\varphi) = z^k H_1(z, \varphi)$ for some other operator H_1. We formally truncate Lyapounov-Schmidt procedure as follows: Define $\psi_k(\lambda, \varphi)$ to be a solution of

$$QG(\lambda, \varphi + \psi_k(\lambda, \varphi)) \equiv 0 \pmod{k}.$$

(This can be done assuming that $G_u(0,0)$ is a Fredholm operator on the space of doubly periodic functions. This was proved by P. Fife [3] for the Bénard problem and by Ravechè and Stuart [5] for the bifurcation models for phase transitions.) Now put

$$F_k(\lambda, \varphi) = PG(\lambda, \varphi + \psi_k(\lambda, \varphi)) .$$

It is clear that $F_k : \Sigma' \to \Sigma'$ and furthermore that it is covariant with respect to the group representation T_g :

$$T_g F_k(\lambda, \varphi) = F_k(\lambda, T_g \varphi) .$$

(The proof of covariance is a modification of the proof of Theorem 4.4.) Furthermore, if φ is Λ-periodic then so is $F_k(\lambda, \varphi)$, and $F_k(\lambda, \varphi)$ coincides up to order k with the bifurcation equations for Λ-periodic functions.

Since F_k is a polynomial in φ we can expand it into linear, quadratic, cubic, ... terms. Say we are interested in the cubic term, (we take $k \geq 3$) and let us write it as $N(\lambda, \varphi)$ where $N(\lambda, z\varphi) = z^3 N(\lambda, \varphi)$. By polarization, N can be written as a trilinear mapping. Suppressing the λ dependence, we thus arrive at a mapping $N(u, v, w)$ which is completely symmetric in u, v, w and which is covariant with respect to the entire group of rigid motions. That is,

$$T_g N(u, v, w) = N(T_g u, T_g v, T_g w)$$

where g is a rigid motion, and

$$N(u_1, u_2, u_3) = N(u_{\sigma(1)}, u_{\sigma(2)}, u_{\sigma(3)})$$

where σ is any permutation of S_3.

Lemma 6.5. Let $u_1, u_2, u_3 \in \mathcal{E}$ and let $\underline{\omega}_1, \underline{\omega}_2, \underline{\omega}_3$ be vectors in $\mathbb{R}^2$. There is a unique operator

$$\hat{N}(u_1, u_2, u_3; \underline{\omega}_1, \underline{\omega}_2, \underline{\omega}_3)$$

from $(\mathcal{E} \times \mathbb{R}^2)^{\otimes 3}$ to $\mathcal{E}$ such that

1) $N(u_1 e^{i<\underline{\omega}_1, \underline{x}>}, u_2 e^{i<\underline{\omega}_2, \underline{x}>}, u_3 e^{i<\underline{\omega}_3, \underline{x}>})$

$$= e^{i<\underline{\omega}_1 + \underline{\omega}_2 + \underline{\omega}_3, \underline{x}>} \hat{N}(u_1, u_2, u_3; \underline{\omega}_1, \underline{\omega}_2, \underline{\omega}_3)$$

2) $S_r(\hat{N}(u_1, u_2, u_3; \underline{\omega}_1, \underline{\omega}_2, \underline{\omega}_3)) = \hat{N}(S_r u_1, S_r u_2, S_r u_3; r\underline{\omega}_1, r\underline{\omega}_2, r\underline{\omega}_3)$

<u>for any rotation</u> r.

3) $\hat{N}(u_1, u_2, u_3; \underline{\omega}_1, \underline{\omega}_2, \underline{\omega}_3) = \hat{N}(u_{\sigma(1)}, u_{\sigma(2)}, u_{\sigma(3)}; \underline{\omega}_{\sigma(1)}, \underline{\omega}_{\sigma(2)}, \underline{\omega}_{\sigma(3)})$

<u>for any</u> $\sigma \in S_3$.

<u>Proof.</u> Actually, the result holds in the case of multilinear operators of any degree. For a linear, Euclidean covariant operator N,

$$T_a N(ve^{i<\underline{\omega}, \cdot>})(\underline{x}) = N(ve^{i<\underline{\omega}, \cdot>})(\underline{x}+\underline{a})$$

$$= N(ve^{i<\underline{\omega}, \cdot + \underline{a}>})(\underline{x})$$

$$= e^{i<\underline{\omega}, \underline{a}>} N(ve^{i<\underline{\omega}, \cdot>})(\underline{x}) .$$

Putting $\underline{x} = 0$ and $\underline{a} = x$ in this equation, we have

$$N(ve^{i<\underline{\omega}, \cdot>})(\underline{x}) = e^{i<\underline{\omega}, \underline{x}>} \hat{N}(v, \underline{\omega}) ,$$

where $\hat{N}(v, \underline{\omega}) = N(ve^{i<\underline{\omega}, \cdot>})(0) = \lim_{x \to 0} N(ve^{i<\underline{\omega}, \cdot>})(\underline{x})$. Since N is also covariant with respect to rotations

$$T_r N(ve^{i<\underline{\omega}, \cdot>})(x) = N(S_r ve^{i<r\omega, \cdot>})(\underline{x})$$

and so $S_r\hat{N}(v, \underline{\omega}) = \hat{N}(S_r v, r\underline{\omega})$. The extension of these arguments to the multilinear case is immediate, and the invariance under the symmetric group is a triviality.

To obtain the bifurcation mapping for any given lattice, choose φ of the form

$$\varphi = z_1 v_1 e^{i<\underline{\omega}_1, \underline{x}>} + \ldots + z_n v_n e^{i<\underline{\omega}_n, \underline{x}>}$$

where $\underline{\omega}_1, \ldots, \underline{\omega}_n$ lie in the lattice and $v_j e^{i<\underline{\omega}_j, \underline{x}>} \in \mathcal{N}$. Then

$$N(\varphi) = \sum_{\underline{\omega}_j + \underline{\omega}_k + \underline{\omega}_\ell = \underline{\omega}_r} z_j z_k z_\ell \hat{N}(v_j, v_k, v_\ell ; \underline{\omega}_j, \underline{\omega}_k, \underline{\omega}_\ell) e^{i<\underline{\omega}_r, \underline{x}>} .$$

Here the vector $\underline{\omega}_r$ lies on the circle of vectors of critical wave-length, so $|\underline{\omega}_r| = \omega_c$. Since the above expression belongs to the kernel $\mathcal{N}$ we must have

$$\hat{N}(v_j, v_k, v_\ell ; \underline{\omega}_j, \underline{\omega}_k, \underline{\omega}_\ell) = q_r(v_j, v_k, v_\ell ; \underline{\omega}_j, \underline{\omega}_k, \underline{\omega}_\ell) v_r$$

where the scalar quantity q_r is given by

$$q_r(v_j, v_k, v_\ell ; \underline{\omega}_j, \underline{\omega}_k, \underline{\omega}_\ell) = <\hat{N}(v_j, v_k, v_\ell ; \underline{\omega}_j, \underline{\omega}_k, \underline{\omega}_\ell), v_r^* > .$$

Here $< , >$ denotes the bilinear pairing between $\mathcal{B}_1$ and its dual; and v_r^* is the adjoint null function, viz. $L_0^* v_r^* e^{i<\underline{\omega}_r, \underline{x}>} = 0$.

Since the full bifurcation equations can be generated by permutations of the variables $z_1, z_2, \ldots$ it is sufficient to calculate

$F_1(z_1, z_2, \ldots)$ -- that is, we need only compute the term

$$\sum_{\underline{\omega}_j+\underline{\omega}_k+\underline{\omega}_\ell} z_j z_k z_\ell \, q_1(v_j, v_k, v_\ell; \underline{\omega}_j, \underline{\omega}_k, \underline{\omega}_\ell) v_1 e^{i<\underline{\omega}_1, \underline{x}>} .$$

Given a monomial $z_j z_k z_\ell$ the number of times it appears in the sum above is equal to the number of distinct solutions of the equation

$$\underline{\omega}_j + \underline{\omega}_k + \underline{\omega}_\ell = \underline{\omega}_1 . \tag{6.11}$$

All solutions of this equation are obtained by setting one of the $\underline{\omega}_j, \underline{\omega}_k$ or $\underline{\omega}_\ell$ equal to $\underline{\omega}_1$ and requiring the other two to sume to zero. There are two types of solutions

3 solutions where $\underline{\omega}_1$ appears twice

6 solutions where $\underline{\omega}_1$ appears once.

(Remark: In the hexagonal case Λ_6 there are actually twelve solutions of (6.11) where $\underline{\omega}_1$ appears once; but the number is cut in half when we symmetrize, that is, when we combine the terms $z_1 z_2 z_5$ and $z_1 z_3 z_6$.)

It remains to compute the expression

$$q_1(v_j, v_k, v_\ell, \underline{\omega}_j, \underline{\omega}_k, \underline{\omega}_\ell) = <\hat{N}(v_j, v_k, v_\ell, \underline{\omega}_j, \underline{\omega}_k, \underline{\omega}_\ell), \underline{v}_1^*) \tag{6.12}$$

whenever (6.12) holds. It suffices to choose $\underline{\omega}_j = \underline{\omega}_1$ and to set $\underline{\omega}_k + \underline{\omega}_\ell = 0$. Let $\underline{\omega}_k$ make an angle θ with $\underline{\omega}_1$: $0 < \theta \leq \frac{\pi}{2}$.

Then we are to calculate

$$q(\theta) = q(v_1, S_\theta v_1, S_{\theta+\pi} v_1; 0, \theta, \theta+\pi)$$

where $0, \theta$, and $\theta+\pi$ are the arguments of the vectors $\underline{\omega}_1$, $\underline{\omega}_j$, and $\underline{\omega}_k = -\underline{\omega}_j$.

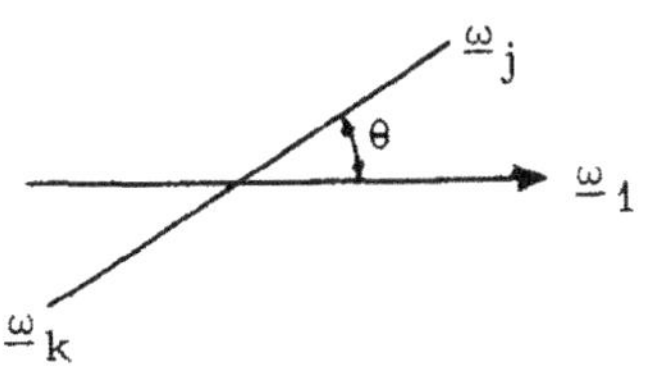

Because of the symmetry of the problem under reflections, $q(\theta) = q(-\theta)$. Furthermore, if we rotate the antipodal vectors $\underline{\omega}_k$ and $\underline{\omega}_j$ through 180° we arrive at the same geometric configuration, but with $\underline{v}_j, \underline{v}_k$ and $\underline{\omega}_j$, $\underline{\omega}_k$ interchanged. By the complete symmetry of $\hat{N}$ we must obtain the same result. In other words, $q(\theta) = q(\theta+\pi)$, and it follows that

$$q(\theta) = A_0 + A_2 \cos 2\theta + A_4 \cos 4\theta + \dots .$$

Our only remaining task is to show the relationship between the coefficients c, d and the lattice function. From the expressions (2.2) and Figure 2.1 we see that c is always the coefficient of a monomial $z_j z_k z_\ell$ where z_1 appears twice and so $c = 3q(0)$. On the other hand, d is always the coefficient of a monomial $z_j z_k z_\ell$ where z_1 appears only once. In the case Λ_2 or Λ_4 there is only one such term, namely $z_1 z_2 z_4$, and it appears 6 times. In the case Λ_6 we have the combined term $z_1(z_2z_5 + z_3z_6)$, of which there are only 6 permutations. (We need count only the permutations of $z_1z_2z_5$). Thus in either case $d = 6q(\theta)$, where θ is the angle between $\underline{\omega}_1$ and $\underline{\omega}_2$. This completes the proof of Theorem 6.4.

It is not possible, in general, to draw any further conclusions about the structure of the lattice function $q(\theta)$. Depending on the internal structure of the equations q may be constant or may contain higher order terms. Let us discuss here the structure of covariant operators and examine the implications for general systems.

If L is a linear translation invariant operator then by the proof of Lemma 6.5,

$$Lve^{i<\underline{\omega},\underline{x}>} = (\hat{L}(\underline{\omega})v)e^{i<\underline{\omega},\underline{x}>} .$$

If L is also covariant with respect to rotations then

$$S_g\hat{L}(\underline{\omega}) = \hat{L}(g\underline{\omega})S_g ,$$

and if S_g happens to be the identity representation, then we have

$$\hat{L}(\underline{\omega}) = \hat{L}(g\underline{\omega}).$$

In this case $\hat{L}(\underline{\omega})$ depends only on the invariant $|\underline{\omega}|$. Examples of such operators are

$$\Delta_2 = \frac{\partial^2}{\partial x^2} + \frac{\partial^2}{\partial y^2} \quad \text{and} \quad \Delta = \frac{\partial^2}{\partial x^2} + \frac{\partial^2}{\partial y^2} + \frac{\partial^2}{\partial z^2} .$$

Similarly, if $N(u, v)$ is a bilinear covariant operator then

$$N(ue^{i<\underline{\omega}, \underline{x}>}, ve^{i<\underline{\sigma}, \underline{x}>}) = \hat{N}(u, v; \underline{\omega}, \underline{\sigma})e^{i<\underline{\omega}+\underline{\sigma}, \underline{x}>}$$

where

$$S_g\hat{N}(u, v; \underline{\omega}, \underline{\sigma}) = \hat{N}(S_gu, S_gv; g\underline{\omega}, g\underline{\sigma}) .$$

In the special case where S_g is the identity representation and $\xi = \mathbb{C}$, u and v are simply scalars and N is a function of the invariants $|\underline{\omega}|$, $|\underline{\sigma}|$, and $\underline{\omega} \cdot \underline{\sigma}$. Examples of such operators are

$$\Delta u \Delta v \qquad \hat{N} = |\underline{\omega}|^2 |\underline{\sigma}|^2$$

$$\nabla u \cdot \nabla v \qquad \hat{N} = \underline{\omega} \cdot \underline{\sigma}$$

$$\Delta^k uv \qquad \hat{N} = |\underline{\omega}+\underline{\sigma}|^{2k} .$$

It is obvious that there is a one-one correspondence between covariant partial differential operators and polynomials in the invariants $|\underline{\omega}|$, $|\underline{\sigma}|$, and $\underline{\omega} \cdot \underline{\sigma}$.

Definition. Let N be an n-linear covariant operator whose polynomial $\hat{N}$ is of degree κ in the dot products $\underline{\omega}_i \cdot \underline{\omega}_j$, $1 \le i < j \le n$. Then we say $\hat{N}$ is of weight κ.

For example, $(\Delta u)(\Delta v)$ has weight zero, and $\nabla u \cdot \nabla v$ has weight one, while $\Delta^2 uv$ has weight 2.

The highest term appearing in the lattice function is determined by the weight of the nonlinear operators involved. Consider, for example, the equation

$$(\Delta^3 - R\Delta_2)u = \alpha u^2 + \beta u^3$$

on the layer $-\infty < x, y < \infty$, $0 < z < d$. All nonlinear operations have weight zero, as does the inverse operator $(\Delta^3 - R\Delta_2)^{-1}$. Therefore the lattice function $q(\theta)$ is constant. The consequence of this fact, as we shall see below, is that rolls are the only stable bifurcating solutions.

In Busses' analysis [1] of cellular convection he notes that in the limit of infinite Prandtl number the Boussinesq equations may be approximated by the following system

$$\Delta\Delta v - \theta = -\gamma(2z\theta - \frac{\theta^2}{R})$$

$$\Delta\theta - R\Delta_2 v = (\delta_j v)(\partial_j \theta)$$

where

$$\Delta_2 = \frac{\partial^2}{\partial x^2} + \frac{\partial^2}{\partial y^2}$$

$$\Delta = \Delta_2 + \partial_z^2$$

$$\delta_j = \frac{\partial}{\partial x_j}$$

$$(\delta_1, \delta_2, \delta_3) = (\frac{\partial^2}{\partial x \partial z}, \frac{\partial^2}{\partial y \partial z}, -\Delta_2) .$$

The quadratic operator $(\delta_j v)(\partial_j \theta)$ is of weight 1. If one traces through the Lyapounov-Schmidt procedure he sees that the lattice function in this case is of weight at most two.

Therefore, the internal structure of the equations in the convection problems allows the possibility of a larger solution structure at the bifurcation point. In cases where it is possible to show that $q(\theta)$ is constant or that q has weight two, that is, that $q(\theta) = A_0 + A_2 \cos 2\theta$, it is a simple matter to translate the conditions in Theorem 6.3 into conditions on A_0 and A_2. We then have

Theorem 6.6 Suppose the lattice function has the form

$$q(\theta) = A \cos 2\theta + B.$$

Then the following pattern selection diagram holds.

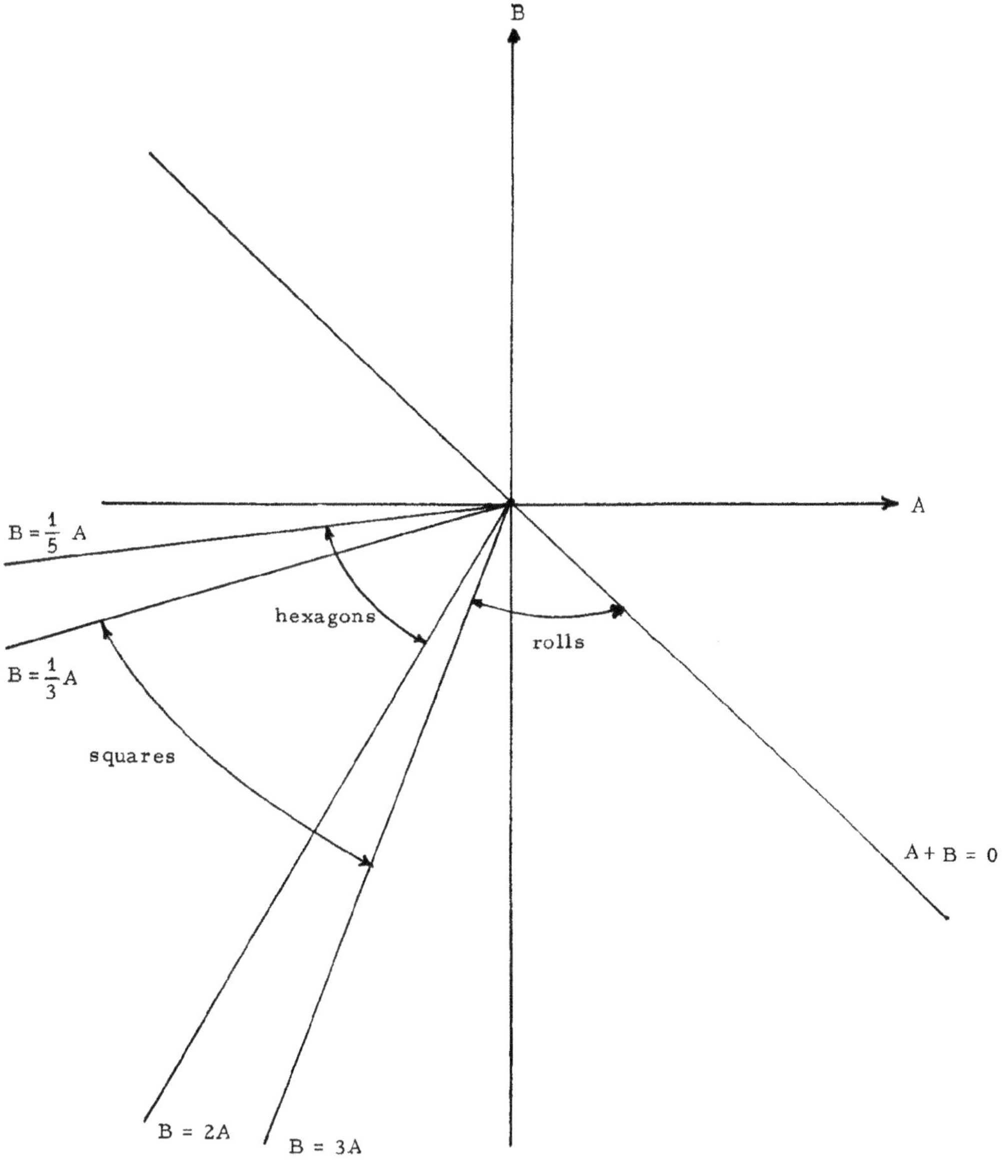
B
A
B = 1/5 A
hexagons
rolls
B = 1/3 A
squares
A + B = 0
B = 2A
B = 3A

In the sector which lies between the lines $B = 3A$ and $A + B = 0$, rolls are the only possible doubly periodic solutions which can be stable. That is, if for a given set of external physical conditions the pair (A, B) lies in this sector, then rolls are the only solutions which can be stable when tested against all disturbances in the three lattice classes. In the sector $2A < B < A/5 < 0$ hexagons may be stable while rolls are unstable. Squares may be stable in the sector $3A < B < A/3$. The hexagons are tested only against disturbances in the hexagonal lattice -- not against disturbances in the other lattices. So far, no mathematical methods are known for making this stability comparison (solutions in one lattice against disturbances in another). There is an overlap in the diagram where squares, rectangles, and hexagons could be similtaneously stable. Thus rolls are the only cellular solutions which are uniquely selected on the basis of a formal stability analysis relative to disturbances within all lattice classes. None of the other patterns are uniquely selected on this basis. Of course, the possibility is open that some of the other patterns are uniquely selected if one tests their stability against still larger clases of disturbances. This would entail a apectral analysis of a linear operator in the plane with doubly periodic coefficients -- a two-dimensional analogue of Hill's equation. The following is immediate:

Theorem 6.7. If the lattice function is constant $(A = 0)$, then rolls are the only possible stable supercritical class of cellular solutions: hexagons, rectangles, and squares, though they exist, are unstable.

Theorem 6.7 applies, in particular, to the case of scalar equations (that is, when $\mathcal{E}$ is just the field of complex numbers). Then rolls are the only stable supercritical solutions in the vicinity of the bifurcation point. These remarks apply, for example, to the bifurcation models for phase transitions investigated by Raveche and Stuart [5], Kozak, Rice, and Weeks [4]. They do not apply to convection problems since the Boussinesq equations do not transform as scalar equations.

6.4. Bifurcation from Rotationally Invariant States

Numerous problems in classical mechanics lead to the breaking of rotational symmetry. For example, the onset of convection in a spherical mass or the breaking of a perfectly uniform spherical shell lead to a bifurcation which breaks complete rotational symmetry. In such cases one is led to investigate bifurcation equations which are covariant with respect to some representation D^{ℓ} of $O(3)$. The methods for deriving the bifurcation equations from the Lie algebra of infinitesimal generators were given in Chapter 5.7.

So far the general solutions of the reduced bifurcation equations for general ℓ are not known. Busse [2] has given a list of special known solutions for even ℓ. In that case the reduced bifurcation equations take the form

$$\lambda z_m = \sum_{m_1+m_2=m} \begin{pmatrix} \ell & \ell & \ell \\ m_1 & m_2 & -m \end{pmatrix} (-1)^m z_{m_1} z_{m_2}$$

where $\begin{pmatrix} \ell & \ell & \ell \\ m_1 & m_2 & m_3 \end{pmatrix}$ are the Wigner 3-j coefficients for the rotation group.

Busse's solutions for even ℓ belong to one of two classes (besides the axisymmetric solutions):

i) $z_0 \neq 0$, $z_n, z_{2n} \neq 0$ $\qquad \frac{1}{3}\ell < n \leq \frac{1}{2}\ell$

z_m otherwise

ii) $z_0 \neq 0$, $z_n \neq 0$ $\qquad$ for a single $n > \frac{\ell}{2}$

$z_m = 0$ $\qquad$ otherwise .

Let us discuss the solutions of the reduced bifurcation equations for $\ell = 2, 4$. We saw in Chapter 5. 7 that the bifurcation equations possessed a gradient structure. Let us use this fact to reformulate the problem for $\ell = 2$. Corresponding to the Clebsch-Gordon series

$$D^1 \otimes D^1 = D^2 \oplus D^1 \oplus D^0 \qquad (6.15)$$

we have the representation

$$U_g A = D^1(g) A D^1(g^{-1}),$$

where A is a 3×3 matrix. U_g is a unitary representation on the vector space of 3×3 matrices with respect to the inner product

$$(A, B) = \frac{1}{2} \operatorname{tr} AB^* .$$

The third-order invariant (there is only one since $D^\ell \otimes D^\ell \otimes D^\ell$ contains D^0 only once) is

$$p(A) = \frac{1}{3} \operatorname{tr} A^2 A^*$$

and this becomes $p(A) = \frac{1}{3} \operatorname{tr} A^3$ if A is Hermitian symmetric. The highest weight space, the one that transforms like D^2 in (6.15), consists of symmetric tensors; so we may restrict ourselves to Hermitian symmetric matrices and rephrase our variational problem as

$$\operatorname{Min} \frac{1}{3} \operatorname{tr} A^3$$

subject to the constraints $\frac{1}{3} \operatorname{tr} A^2 = 1$, $\operatorname{tr} AB_j = 0$, where the B_j are symmetric matrices which lie in the lower weight invariant subspaces. In particular, the symmetric matrix which transforms according to D^0 must be the identity matrix; for if $D^1(g)BD^1(g^{-1}) = B$ then by Schur's lemma, B is a scalar multiple of the identity. Therefore, we get the constraints

$$\frac{1}{2}\operatorname{tr} A^2 = 1 , \quad \operatorname{tr} A = 0.$$

The Euler-Lagrange equations for this variational problem are

$$A^2 = \lambda A + \gamma I \tag{6.16}$$

These equations may be completely solved as follows. Taking the trace and using the constraints we get $\gamma = 2/3$. We then choose a rotation g so that $D^1(g)AD^1(g^{-1})$ is diagonal (we can do this since $D^1(g)$ ranges over all orthogonal matrices as g ranges over $O(3)$). Since (6.16) is covariant, we can assume A is diagonal; then (6.16) is equivalent to

$$\mu_i^2 = \lambda\mu_i + \frac{2}{3} \qquad i = 1, 2, 3 \tag{6.17}$$

where μ_i are the eigenvalues of A. The constraints are

$$\mu_1^2 + \mu_2^2 + \mu_3^3 = 2 , \quad \mu_1 + \mu_2 + \mu_3 = 0 . \tag{6.18}$$

There are two sets of solutions to these equations, namely,

$$\lambda = -\frac{1}{\sqrt{3}} \quad , \qquad A = \begin{pmatrix} \frac{1}{\sqrt{3}} & 0 & 0 \\ 0 & \frac{1}{\sqrt{3}} & 0 \\ 0 & 0 & -\frac{2}{\sqrt{3}} \end{pmatrix}$$

and

$$\lambda = \frac{1}{\sqrt{3}} \quad , \qquad A = \begin{pmatrix} -\frac{1}{\sqrt{3}} & 0 & 0 \\ 0 & -\frac{1}{\sqrt{3}} & 0 \\ 0 & 0 & \frac{2}{\sqrt{3}} \end{pmatrix}$$

The order of the eigenvalues on the diagonal of A is immaterial, for any permutation of the diagonal entries of A produces a point on the same orbit. Indeed, any such permutation is accomplished by the operation PAP^{-1}, where P is a permutation matrix, and such a P is an element of $O(3)$. One of the orbits above gives the maximum of the functional $\frac{1}{3}\operatorname{tr} A^3$ on the sphere $\frac{1}{2}\operatorname{tr} A^2 = 1$; the other orbit gives the minimum. The isotropy subgroup in each case is $O(2)$ (rotations which leave $\begin{pmatrix} 0 \\ 0 \\ 1 \end{pmatrix}$ invariant), so each extremal is axisymmetric. (This procedure for solving the reduced bifurcation equations for $\ell = 2$ was developed jointly with Professor L. Green at the School of Mathematics at the University of Minnesota. The method is closely related to the ideas of L. Michel discussed in Chapter I.)

The approach described above, while quite straightforward in the case $\ell = 2$, becomes extremely complicated already in the case $\ell = 4$ and so does not seem to be a practical method for the resolution of the reduced bifurcation equations in the general case. For $\ell = 4$ the quadratic terms in the bifurcation equations are

$$F_4 = \frac{1}{\sqrt{5}} z_4 z_0 - \frac{1}{\sqrt{2}} z_3 z_1 + \frac{3}{2\sqrt{14}} z_2^2$$

$$F_3 = \frac{1}{\sqrt{2}} z_4 z_{-1} - \frac{3}{2\sqrt{5}} z_3 z_0 + \frac{1}{\sqrt{14}} z_2 z_1$$

$$F_2 = \frac{3}{\sqrt{14}} z_4 z_{-2} - \frac{1}{\sqrt{14}} z_3 z_{-1} - \frac{11}{14\sqrt{5}} z_2 z_0 + \frac{3}{7\sqrt{2}} z_1^2$$

$$F_1 = \frac{1}{\sqrt{2}} z_4 z_{-3} + \frac{1}{\sqrt{14}} z_3 z_{-2} - \frac{6}{7\sqrt{2}} z_2 z_{-1} + \frac{9}{9\sqrt{20}} z_1 z_0$$

$$F_0 = \frac{1}{\sqrt{5}} z_4 z_{-4} + \frac{3}{2\sqrt{5}} z_3 z_{-3} - \frac{11}{14\sqrt{5}} z_2 z_{-2} - \frac{9}{14\sqrt{5}} z_1 z_{-1} + \frac{9}{14\sqrt{5}} z_0^2$$

The remaining polynomials are found from those above by the relationship $F_{-m}(z_{-4},\ldots,z_4) = (-1)^m F_m(z_{-4},\ldots,z_4)$. There are many possible solutions to the bifurcation equations in this case. Busse has found two special solutions

1) Axisymmetric solutions: $z_{\pm 1} = \ldots = z_{\pm 4} = 0, \quad z_0 \neq 0.$

2) Octahedral solutions: $z_4 = z_{-4} = \frac{5}{\sqrt{14}}, \quad z_0 = \sqrt{5},$

$$z_{\pm 1} = z_{\pm 2} = z_{\pm 3} = 0 .$$

Busse conjectures, on the basis of numerical work, that the octahedral solution is the one which maximizes the third-order invariant.

An analysis of the Jacobian shows that the axisymmetric solution is a saddle point and that the octahedral solution is a candidate for the maximum. The Jacobian at the octahedral solution is (we set $\lambda = 1$ and drop the normalization condition $|z| = 1$; the results are affected only by a possible change of scale):

$$J - I = \begin{bmatrix} 0 & 0 & 0 & 0 & \sqrt{5/14} & 0 & 0 & 0 & 0 \\ 0 & -5/2 & 0 & 0 & 0 & 5/2\sqrt{7} & 0 & 0 & 0 \\ 0 & 0 & -25/14 & 0 & 0 & 0 & 15/14 & 0 & 0 \\ 0 & 0 & 0 & -5/14 & 0 & 0 & 0 & 5/2\sqrt{7} & 0 \\ \sqrt{5/14} & 0 & 0 & 0 & 2/7 & 0 & 0 & 0 & \sqrt{5/14} \\ 0 & 5/2\sqrt{7} & 0 & 0 & 0 & -5/14 & 0 & 0 & 0 \\ 0 & 0 & 15/14 & 0 & 0 & 0 & -25/14 & 0 & 0 \\ 0 & 0 & 0 & 5/2\sqrt{7} & 0 & 0 & 0 & -5/2 & 0 \\ 0 & 0 & 0 & 0 & \sqrt{5/14} & 0 & 0 & 0 & 0 \end{bmatrix}$$

The eigenvalues of this matrix can be determined by restricting the matrix to certain invariant subspaces (determined by inspection) as follows. Let $\underline{e}_i$ denote the column vector with a one in the ith row and zeroes everywhere else. The subspaces $\{a\underline{e}_3 + b\underline{e}_7\}$, $\{a\underline{e}_1 + b\underline{e}_5 + c\underline{e}_9\}$, $\{a\underline{e}_2 + b\underline{e}_6\}$, $\{a\underline{e}_4 + b\underline{e}_8\}$ are all invariant, and one has to calculate the eigenvalues of a 3×3 matrix at most. The complete set of eigenvalues is

$$\{0, 0, 0, -\frac{20}{7}, -\frac{20}{7}, -\frac{20}{7}, -\frac{5}{7}, -\frac{5}{7}, 1\}.$$

This set of eigenvalues indicates that the (transcritical) branch of solutions has one unstable subcritical mode. In fact, if the reduced bifurcation equations are $\lambda z + B(z, z) = 0$ then the appropriate scaling would be

$$\lambda = -\epsilon \ , \quad z = \epsilon\xi \ ;$$

this would lead to the Jacobian $-J + I$ given above, and $\epsilon > 0$ would lead to subcritical branching; such a subcritical branch would have three neutral modes, five stable modes, and one unstable mode. Since only one eigenvalue is positive the octahedral solution (it has octahedral symmetry) is a possible candidate for the maximum of the external problem. The axisymmetric solution above, however, is definitely a saddle point of the variational problem; the eigenvalues of its Jacobian are

$$\{0, 0, \frac{20}{9}, \frac{20}{9}, \frac{10}{3}, \frac{10}{3}, -\frac{5}{9}, -\frac{5}{9}, -1\}.$$

The relationship between the stability of the bifurcating solutions and the external problem will be discussed in the following section (Theorem 6.10).

Two open problems in the bifurcation problems here are

(1) Determine, for each ℓ, the isotropy subgroup of the extremal solution of the variational problem.

(2) Extend each solution of the reduced bifurcation equations to a solution of the full bifurcation equations; (the difficulty here lies in that the implicit function theorem cannot be directly applied since the solutions always appear in two or three dimensional orbits.)

6.5 Variational Methods in Bifurcation Theory

Let $\mathcal{H}$ be a Hilbert space with inner product $(\,,\,)$. A mapping $B(u)$ is the gradient of a functional $\mathcal{F}(u)$ if

$$\mathcal{F}(u + \epsilon h) = \mathcal{F}(u) + \epsilon (B(u), h) + O(\epsilon^2).$$

Let $B(u, v)$ be a symmetric bilinear mapping on $\mathcal{H}$, and let $\mathcal{F}(u) = \frac{1}{3} B(u, u), u)$. Then

Lemma 6.8. *Let* B, $\mathcal{J}$, *and* $\mathcal{H}$ *be as above. Then* B *is the gradient of* $\mathcal{J}$ *if and only if the trilinear form* (B(u, v), w) *is completely symmetric.*

Proof. If (B(u, v), w) is completely symmetric then

$$\mathcal{J}(x+\epsilon h) = \mathcal{J}(x) + \frac{\epsilon}{3}[(B(h,x),x) + (B(x,h),x) + (B(x,x),h)] + O(\epsilon^2)$$

$$= \mathcal{J}(x) + \epsilon (B(x,x),h) + O(\epsilon^2) \; ;$$

Consequently, B(x, x) is the gradient of $\mathcal{J}(x)$.

Conversely, by the symmetry of B,

$$\mathcal{J}(x+\epsilon h) = \mathcal{J}(x) + \epsilon[(B(x,x),h) + 2(B(x,h),x)] + O(\epsilon^2);$$

whereas, if B is the gradient of $\mathcal{J}$ then

$$\mathcal{J}(x+\epsilon h) = \mathcal{J}(x) + \epsilon(B(x,x),h) + O(\epsilon^2).$$

Comparing terms of order ϵ we have (B(x, x), h) = (B(x, h), x) for all vectors x and h. Replacing x by x + y in this identity and using the symmetry of B we obtain the identity

$$(B(x,y),h) = \frac{1}{2}[(B(x,h),y) + (B(y,h),x)].$$

Interchanging x and h in the above identity we get

$$
\begin{aligned}
(B(h,y),x) &= \frac{1}{2}[(B(h,x),y) + (B(y,x),h] \\
&= \frac{1}{2}[(B(h,x),y) + \frac{1}{2}\{(B(x,h),y) + (B(y,h),x)]] \\
&= \frac{3}{4}(B(x,h),y) + \frac{1}{4}(B(y,h),x) \; ;
\end{aligned}
$$

hence $(B(y,h),x) = (B(h,x),y) = (B(x,h),y)$ and the trilinear form $(B(x,y),h)$ is completely symmetric.

The following is an immediate consequence of the symmetry of the 3-j symbols.

Theorem 6.9. Let the bilinear mapping $B(u,v)$ be covariant with respect to the representation D^{ℓ} of SO(3). Then the trilinear form $(B(u,v),w)$ is symmetric if ℓ is even and skew if ℓ is odd.

Proof. The trilinear form $(B(u,v),w)$ is a third-order invariant of the tensor product $D^{\ell} \otimes D^{\ell} \otimes D^{\ell} = \Gamma$. Now $D^{\ell} \otimes D^{\ell}$ contains D^{ℓ} exactly once, and D^{0} arises in Γ due to the tensor product $D^{\ell} \otimes D^{\ell}$. Therefore, Γ contains the identity representation precisely once. If $F(u_1, u_2, u_3)$ is an invariant of Γ then so is $F(u_{\sigma(1)}, u_{\sigma(2)}, u_{\sigma(3)})$ for any permutation σ. So F must transform according to a one-dimensional representation of S_3. Therefore F is either symmetric or skew. Furthermore, if

$F(u,v,w) = (B(u,v),w)$ where B is a covariant of D^ℓ, then F must be skew when ℓ is odd and symmetric when ℓ is even. For B is skew for odd ℓ and symmetric for even ℓ. This can be seen by examining the Clebsch-Gordon series

$$D^\ell \otimes D^\ell = D^{2\ell} \oplus D^{2\ell-1} \oplus \ldots \oplus D^\ell \oplus \ldots \oplus D^0.$$

The first vector space, the one that transforms like $D^{2\ell}$, consists of symmetric tensors, the next of skew tensors, and so forth. The covariant mapping B corresponds to a set of tensors which transform among themselves as D^ℓ.

As a consequence of Theorem 6.9 the quadratic term in the bifurcation equations vanishes if ℓ is odd and is a gradient if ℓ is even.

Next let us analyze the relationship between the stability of the bifurcating solutions and the extremal problem.

<u>Theorem 6.10.</u> <u>Let</u> B <u>be the gradient of</u> $\mathfrak{F}$ <u>as above and suppose the reduced bifurcation equations take the form</u>

$$\sigma v + B(u,v) = 0.$$

<u>These are the Euler-Lagrange equations for the variational problem</u>

$$\max_{(v,v)=1} \mathfrak{F}(v) .$$

The Jacobian of the reduced equations is $J_\sigma(v)x = \sigma x + 2B(v,x)$. At the maximum point v, one eigenvalue of $J_\sigma(v)$ is negative and the rest are positive. Accordingly, from Theorem 4.3 it follows that the corresponding branch, which in this case is transcritical, has one unstable subcritical mode.

This situation occurs often in classical mechanics. We have already seen it in the case of the bifurcation of solutions in the hexagonal lattice; and it occurs again when rotational symmetry is broken if the kernel $\mathcal{N}$ transforms according to an even representation of O(3) (D^ℓ, even). If the subcritical branch bends back and regains stability, as pictured in Figure 6.3, then there is the possibility of jump transitions. In classical nonlinear oscillation theory, this situation is known as "hard excitation", while in buckling theory it is called a "snap-through instability". As we see, such discontinuous phenomena are tied up with the symmetries of the problem.

Proof of Theorem 6.10. We calculate the second variation at an extremal v. Let $x(t)$ be a curve on the unit sphere such that $x(0) = v$. Then if $\mathcal{J}$ attains a maximum at v,

$$\frac{d^2}{dt^2}\,\mathcal{J}(x(t)) = (B(v,v),\ddot{x}) + 2(B(v,\dot{x}),\dot{x}) \le 0$$

and

$$\frac{d^2}{dt^2}\frac{1}{2}(x(t), x(t)) = (v, \ddot{x}) + (\dot{x}, \dot{x}) = 0 .$$

At v, $\sigma v + B(v, v) = 0$, so

$$-\sigma(v, \ddot{x}) + 2(B(v, \dot{x}), \dot{x}) \leq 0,$$

$$(\sigma\dot{x} + 2B(v, \dot{x}), \dot{x}) \leq 0$$

for all tangent vectors $\dot{x}$ at v. Consequently, $(J\sigma(v)\dot{x}, \dot{x}) \leq 0$ for all such tangent vectors $\dot{x}$. Furthermore, $J\sigma(v)$ leaves the tangent plane at v invariant. In fact, the equation $(v, \dot{x}) = 0$ describes the tangent plane at v and also $(B(v, \dot{x}), v) = (B(v, v), \dot{x}) = -\sigma(v, \dot{x}) = 0$. Therefore, if $(v, \dot{x}) = 0$ then $(J\sigma(v)\dot{x}, v) = 0$ as well, and the tangent plane at v is preserved. Since $J\sigma(v)$ maps the tangent plane to itself, the inequality $(J\sigma(v)\dot{x}, \dot{x}) \leq 0$ tells us that $J\sigma$ is negative semi-definite at a local maximum v. The normal vector to the tangent plane at v is v itself, and $J\sigma(v)v = \sigma v + 2B(v, v) = B(v, v) = -\sigma v$, so the remaining eigenvalue of $J\sigma$ is $-\sigma$. Since at an external v

$$-\sigma = \frac{(B(v, v), v)}{(v, v)}$$

the eigenvalue $-\sigma$ is positive at a positive maximum of $(B(v, v), v)$.

If the original bifurcation equations are

$$\lambda w + B(w, w) + \ldots$$

then by scaling $\lambda = \epsilon\sigma$ and $w = \epsilon v$, dividing by ϵ^2, and letting $\epsilon \to 0$ we arrive at the reduced equations $\sigma v + B(v, v) = 0$. When we extend the solution corresponding to the maximum of the variational problem, $\sigma < 0$ (as we saw above), and so $\epsilon > 0$ corresponds to subcritical bifurcation. Accordingly, negative eigenvalues of $J\sigma(v)$ give rise to stable subcritical modes, and positive eigenvalues of $J\sigma(v)$ give rise to unstable subcritical modes.

A similar analysis can be carried through in the cubic case, when the reduced bifurcation equations take the form

$$\sigma x - B(x, x, x) = 0.$$

Again, if B is the gradient of the functional $\mathfrak{J}(x) = \frac{1}{4}(B(x, x, x), x)$ then the reduced equations are the Euler-Lagrange equations for the variational problem

$$\underset{(x,x)=1}{\mathrm{Min}} \quad \frac{1}{4}(B(x, x, x), x).$$

It can easily be shown, by the same analysis as before, that the eigenvalues of the Jacobian are always non-positive at a positive minimum of the quartic $\mathfrak{J}(x)$ on the sphere $(x, x) = 1$. Hence at a positive minimum (which does not necessarily exist) we get stable supercritically branching solutions. The bifurcations in this case are always one-sided: supercritical at positive extrema and subcritical at negative extrema. Subcritical solutions are always unstable.

That the reduced bifurcation equations may possess a gradient structure has been observed by Busse in connection with the Bénard problem [1] and the breaking of rotational symmetry [2]. We have seen in this chapter that the gradient structure may be a consequence of the group covariance of the bifurcation equations, and so may hold even if the original problem does not possess a gradient structure. In the case of O(3), the quadratic term vanishes for odd ℓ and is a gradient for even ℓ, as we proved in Theorem 6.9. Accordingly, for even ℓ, the bifurcation is transcritical with one unstable subcritical mode; and the possibility of jump transitions exists. Even for odd ℓ, the reduced bifurcation equations still possess a gradient structure. This was shown to me by Professor L. Michel of the Institut des Hautes Etudes Scientifiques. The proof, however, is more involved and requires a knowledge of the representations of the symmetric group.

REFERENCES

1. F. Busse, "The stability of finite amplitude cellular convection and its relationship to an extremum principle," J. Fluid Mech. 30 (1967), 625-650.

2. F. Busse, "Patterns of convection in spherical shells," J. Fluid Mech. 72 (1975), 67-85.

3. P. Fife, "The Bénard problem for general fluid dynamical equations and remarks on the Boussinesq approximations," Indiana Univ. Math. Jour. 20 (1973).

4. Kozak, Rice, and Weeks, "Analytic approach to the theory of phase transitions," J. Chem. Phys. 52 (1970), p. 2416.

5. Ravechè, H. J. and Stuart, C. A., " Towards a molecular theory of freezing," J. Chem. Phys. 63 (1975), 136-152.

6. Sattinger, D. H., "Group representation theory, bifurcation theory and pattern formation," Jour. Functional. Anal. 28 (1978), 58-101.

7. Sattinger, D. H., "Selection mechanisms for pattern formation," Arch. Rat. Mech. Anal. 66 (1977), 31-42.

8. Sattinger, D. H., "Bifurcation from rotationally invariant states," Jour. Math. Phys. 19 (1978), 1720-1732.

VII

APPENDIX: HOW TO FIND THE SYMMETRY GROUP OF A DIFFERENTIAL EQUATION

Peter J. Olver

Before applying group-theoretic methods to the construction of the bifurcation equations of some system of partial differential equations, it is of course necessary to know a group of symmetries of the equations in question. In this chapter we describe a useful, systematic computational method for finding the group of symmetries of a given system of partial differential equations. This method essentially dates back to the original investigations of Sophus Lie; other modern treatments of this material may be found in references [2] and [5]. The groups under consideration will be local Lie groups transforming both the independent and dependent variables of the differential equations. Thus, we will leave aside any questions on the discrete symmetries of the equation. The reason for this restriction is to take full advantage of the infinitesimal methods available in Lie group theory. For simplicity, we will work in

Euclidean spaces although many of the results hold equally well for differential equations on manifolds. (See [4] for a rigorous exposition.)

1. Local Transformation Groups

Definition 7.1. A local group of transformations acting on $\mathbb{R}^n$ consists of a Lie group G, an open set V, with $\{e\} \times \mathbb{R}^n \subset V \subset G \times \mathbb{R}^n$, and a smooth ($C^\infty$) map $\Phi : V \to \mathbb{R}^n$, satisfying the conditions

i) $\Phi(e, x) = x$ for $x \in \mathbb{R}^n$

ii) $\Phi(g, \Phi(h, x)) = \Phi(g \cdot h, x)$

whenever $g, h \in G$, $x \in \mathbb{R}^n$, and $(h, x), (g, \Phi(h, x)), (g \cdot h, x) \in V$. (In other words, this equation holds whenever both sides make sense.)

If $V = G \times \mathbb{R}^n$, then the group action of G is global. In general, however, for each $x \in \mathbb{R}^n$, only those group elements in a neighborhood of e in G (depending on x) can transform x. Examples of local transformation groups whose action cannot be globalized arise naturally as symmetries of partial differential equations. Note further that the action of the group is not restricted to be linear.

Associated with a local transformation group are its infinitesimal generators. These are vector fields on $\mathbb{R}^n$ defined as follows: Let $\mathfrak{g}$ denote the Lie algebra of G. Given $\alpha \in \mathfrak{g}$, let $\exp(t\alpha)$ be the one-parameter subgroup of G generated by α. The corresponding infinitesimal generator on $\mathbb{R}^n$ is the vector field $\varphi(\alpha)$ whose value at $x \in \mathbb{R}^n$ is

$$\varphi(\alpha)\Big|_x = \frac{d}{dt}\Big|_{t=0} \Phi(\exp(t\alpha), x) . \tag{7.1}$$

If $x = (x_1, \dots, x_n)$ are coordinates on $\mathbb{R}^n$, we shall adopt the differential-geometric notation

$$\vec{v} = \xi^1(x) \frac{\partial}{\partial x_1} + \dots + \xi^n(x) \frac{\partial}{\partial x_n}$$

for vector fields on $\mathbb{R}^n$. Thus if $\Phi(\exp(t\alpha), x) = (\Phi^1(t, x), \dots, \Phi^n(t,x))$, then the infinitesimal generator has coordinate functions $\xi^i(x) = \frac{d}{dt} \Phi^i(t, x)\Big|_{t=0}$. Conversely, given a vector field $\vec{v}$, as above, the one-parameter local group of transformations generated by $\vec{v}$ is found by solving the system of ordinary differential equations

$$dx_i/dt = \xi^i(x) , \quad i = 1, \dots, n ,$$
$$x(0) = x .$$

Vector fields on $\mathbb{R}^n$ can also be viewed as first-order partial differential operators (derivations) which act on smooth functions $F: \mathbb{R}^n \to \mathbb{R}$:

$$\vec{v}F(x) = \xi^1(x)\frac{\partial F}{\partial x_1} + \ldots + \xi^n(x)\frac{\partial F}{\partial x_n} .$$

Given two vector fields $\vec{v}$ and $\vec{w}$, the Lie bracket is the vector field

$$[\vec{v}, \vec{w}] = \vec{v}\,\vec{w} - \vec{w}\,\vec{v} ,$$

where we are viewing the vector fields as derivations. The map φ from the Lie algebra to vector fields defined by (7.1) preserves the Lie bracket:

$$\varphi([\alpha, \beta]) = [\varphi(\alpha), \varphi(\beta)] .$$

Thus the infinitesimal generators of a local transformation group form a finite-dimensional Lie algebra of vector fields on $\mathbb{R}^n$. Conversely, given a finite-dimensional Lie algebra of vector fields on $\mathbb{R}^n$, Frobenius' theorem (cf. [3] or [8]) says that there is a local transformation group whose infinitesimal generators are precisely the vector fields in question. We are thus justified in viewing local transformation groups and Lie algebras of vector fields as equivalent concepts. In practice, to find the symmetry group of a differential equation, the infinitesimal generators will in fact be calculated, this being much easier to accomplish.

Example 7.2. Let $G = \mathbb{R}$ with coordinate t, and consider the following action on $\mathbb{R}^2$

$$\Phi(t; x, y) = \left(\frac{x}{1-ty}, \frac{y}{1-ty} \right) .$$

Here $V = \{(t; x, y): t < y \text{ for } y > 0 \text{ and } t > y \text{ for } y < 0\}$. The reader should check that Φ does satisfy the conditions of Definition 7.1. Also, this action is not global, i.e., cannot be defined for all $t \in \mathbb{R}$. The infinitesimal generator of the group action is

$$\left.\frac{d}{dt}\right|_{t=0} \Phi(t; x, y) = xy \frac{\partial}{\partial x} + y^2 \frac{\partial}{\partial y} .$$

Indeed, as can easily be checked, the coordinate functions satisfy the ordinary differential equations

$$dx/dy = xy \qquad dy/dt = y^2 .$$

In the sequel we will often use the simplified notation of denoting $\Phi(g, x)$ by gx. Also, an element α of the Lie algebra will be identified with the vector field it defines on $\mathbb{R}^n$, and the map φ will be suppressed.

Now, a function $F: \mathbb{R}^n \to \mathbb{R}^m$ is called a G-invariant function if for all $x \in \mathbb{R}^n$, $F(gx) = F(x)$ whenever gx is defined. Similarly, a subset $S \subset \mathbb{R}^n$ is called a G-invariant subset if for every $x \in S$, $gx \in S$ whenever gx is defined. Note that if F is a G-invariant function, all the level sets of F, $\{x : F(x) = c\}$, are G-invariant sets. However, if a set S is a <u>subvariety</u> given by the vanishing of a function, i.e., $S = \{x : F(x) = 0\}$, and S is G-invariant, it does not necessarily follow that F is a G-invariant function. Thus the symmetry group of a single level set of a function (meaning the "largest group of transformations" leaving the level set invariant) will in general contain more symmetries than the symmetry group of the function. The next theorem gives <u>infinitesimal criteria</u> for the invariance of a function or subvariety.

<u>Theorem 7.3.</u> <u>Suppose</u> G <u>is a connected Lie group of transformations acting on</u> $\mathbb{R}^n$, <u>such that for each</u> $x \in \mathbb{R}^n$, $G_x = \{g : gx$ <u>is defined</u>$\}$ <u>is also connected. Let</u> $F : \mathbb{R}^n \to \mathbb{R}^m$ <u>be a differential function whose Jacobian matrix is of maximal rank everywhere.</u>

i) F <u>is a G-invariant function if and only if</u>

$$\alpha F(x) = 0 \tag{7.2}$$

<u>for every infinitesimal generator</u> α <u>of</u> G <u>and every</u> $x \in \mathbb{R}^n$.

ii) <u>The subvariety</u> $S = \{x : F(x) = 0\}$ <u>is G-invariant if and only if (7.2) holds for every infinitesimal generator</u> α <u>and for every</u> $x \in S$.

Proof. First the second statement will be proven. The necessity of (7.2) follows from differentiating the equation

$$F(\exp(t\alpha)x) = 0 \qquad \alpha \in \mathfrak{g}, \ x \in S,$$

with respect to t and setting $t = 0$. To prove sufficiency, if the Jacobian of F at x_0 has maximal rank then by the implicit function theorem we may locally change coordinates so that F has the form

$$F(x_1, \ldots, x_n) = (x_1, \ldots, x_m).$$

Thus $S = \{(0, \ldots, 0, x_{m+1}, \ldots, x_n)\}$. The infinitesimal condition (7.2) implies that α, when restricted to S, has the form $\alpha_{m+1}(x) \frac{\partial}{\partial x_{m+1}} + \ldots + \alpha_n(x) \frac{\partial}{\partial x_n}$, $x \in S$. Then the one-parameter subgroup $\exp(t\alpha)$ obtained by integrating the requisite system of o.d.e.'s obviously leaves S locally invariant. Hence for each $x \in S$, there is a neighborhood $\tilde{N}_x$ of 0 in $\mathfrak{g}$ such that for $\alpha \in \tilde{N}_x$, $\exp(\alpha)x \in S$. Now $\exp: \mathfrak{g} \to G$ maps a sufficiently small neighborhood of 0 in $\mathfrak{g}$ homeomorphically onto a neighborhood of e in G, [8; page 103]. Thus for each $x \in S$, there is a neighborhood N_x of e in G (depending continously on x) such that whenever $g \in N_x$, $gx \in S$. Finally, to show S is G-invariant, given $x \in S$, let $H_x = \{g \in G : gx \text{ is defined and } gx \in S\}$. It is easy to show that H_x is open and also if $g \in \operatorname{clos} H_x$, then $g \cdot x$ is not defined. This implies, by the connectedness of G_x, that $H_x = G_x$ and hence S is indeed G-invariant. To prove part i), it suffices to note that F is invariant if and only if every level set of F is invariant. Therefore part ii) implies part i).

(In the sequel, when G is a connected group of transformations, it will be implicitly assumed that all the G_x's are connected so as to avoid restating this technical condition.)

In practice, if the vector fields $\alpha_i = \sum \alpha_i^k(x)\, \partial/\partial x_k$, $i = 1, \ldots, r$, form a basis of the Lie algebra and $F(x) = (F^1(x), \ldots, F^m(x))$ then the infinitesimal criterion of (7.2) is just

$$\sum_{k=1}^{n} \alpha_i^k(x) \frac{\partial F^j}{\partial x_k} = 0 , \qquad \begin{array}{l} i = 1, \ldots, r , \\ j = 1, \ldots, m . \end{array} \tag{7.2'}$$

In the second case of the theorem, these equations only need hold when $F^1(x) = \ldots = F^m(x) = 0$. Thus determining the invariance of a subvariety under a connected group reduces to a routine verification of condition (7.2'). For nonconnected Lie groups, one must further check that at least one element in each connected component leaves the subvariety invariant.

Example 7.4. Consider the same local group of transformations as in example 7.2. First consider the function $F(x, y) = x/y$. Then applying the infinitesimal generator to F shows

$$(xy\frac{\partial}{\partial x} + y^2\frac{\partial}{\partial y})\frac{x}{y} = 0$$

hence F is an invariant function, as may easily be checked. Therefore the sets $\{(x, y) : x = cy\}$ for any constant c are invariant.

Secondly, consider the function $F'(x, y) = xy$. Then

$$(xy\frac{\partial}{\partial x} + y^2\frac{\partial}{\partial y})xy = 2xy^2 ;$$

so F' is not an invariant function. However, the subvariety $\{(x, y) : xy = 0\}$ is invariant since if $xy = 0$, then $2xy^2 = 0$. This may again be verified directly from the definition of G.

2. Groups and Differential Equations.

Suppose we are considering a system of partial differential equations, S, in p independent variables $(x_1, \ldots, x_p) = x$ and q dependent variables $(u^1, \ldots, u^q) = u$. Let $X = \mathbb{R}^p$, with coordinates x, be the space representing the independent variables, and let $U = \mathbb{R}^q$, with coordinates u represent the dependent variables. The solutions $u = f(x)$ of S will be identified with their graphs; which are certain p-dimensional submanifolds in the cartesian product space $X \times U$. A symmetry group of S will be a local group of transformations G acting on $X \times U$ in such a way that "G transforms solutions of S to other solutions of S." Note that we are allowing arbitrary, nonlinear transformations of both the independent and dependent variables in G.

To proceed rigorously, we must first explain exactly how the group G transforms functions. Given a function $u = f(x)$, defined in a neighborhood N of a point $x_0 \in X$, let $\Gamma = \{(x, f(x)) : x \in N\}$ be the graph of f. If Γ is relatively compact in $X \times U$, then, for g sufficiently close to the identity, the set $g\Gamma = \{g(x,u) : (x,u) \in \Gamma\}$ is defined. The set $g\Gamma$ is not necessarily the graph of some other function. However, since G acts continuously and e leaves Γ unchanged, by possibly shrinking N we can find a neighborhood of e in G such that for every g in this neighborhood, $g\Gamma$ is defined and is the graph of some new function $\tilde{u} = g \cdot f(\tilde{x})$, called the transform of f by g.

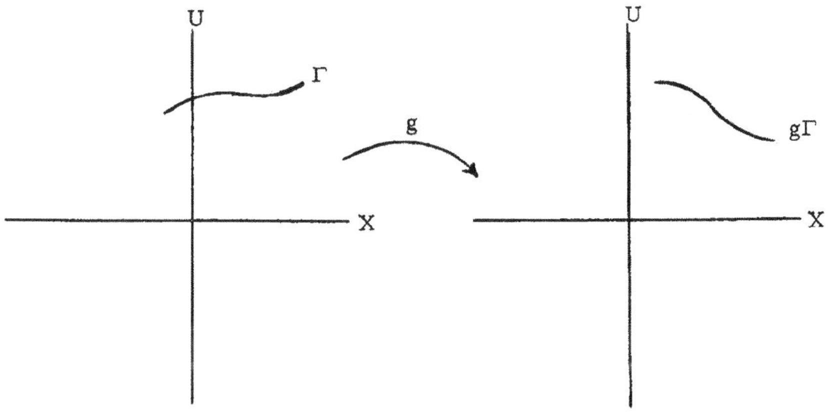

The explicit construction of the transformed function gf follows. Suppose the transformation g is given by

$$g(x, u) = (\Xi_g(x,u), \Psi_g(x, u)) = (\tilde{x}, \tilde{u}).$$

The graph $g\Gamma$ is given by the parametric equations

$$\tilde{x} = \Xi_g(x, f(x)) = \Xi_g \circ (I \times f)(x) ,$$

$$\tilde{u} = \Psi_g(x, f(x)) = \Psi_g \circ (I \times f)(x) .$$

(Here I is the identity function on X.) To find $g \cdot f$, we must eliminate x from these equations. For g sufficiently close to e, using the inverse function theorem we can locally solve for x:

$$x = [\Xi_g \circ (I \times f)]^{-1}(\tilde{x}) .$$

Substitution into the second equation yields

$$g \cdot f = [\Psi_g \circ (I \times f)] \circ [\Xi_g \circ (I \times f)]^{-1} , \tag{7.3}$$

whenever the second factor is invertible.

<u>Example 7.5.</u> Let $p = q = 1$, so $X = U = \mathbb{R}$. Let $G = S_1$ be the rotation group acting on $X \times U \simeq \mathbb{R}^2$, so the transformations in G are given by

$$(\tilde{x}, \tilde{u}) = (x \cos\theta - u \sin\theta,\ x \sin\theta + u \cos\theta)$$

for $0 \le \theta < 2\pi$. Therefore

$$\Xi_\theta(x, u) = x\cos\theta - u\sin\theta ,$$

$$\Psi_\theta(x, u) = x\sin\theta + u\cos\theta .$$

Consider the linear function $f(x) = ax+b$. Note that if θ is sufficiently large, the graph of f will be rotated so that it is vertical and is no longer the graph of a function. Now

$$\tilde{x} = \Xi_\theta(x, f(x)) = x(\cos\theta - a\sin\theta) - b\sin\theta ,$$

$$\tilde{u} = \Psi_\theta(x, f(x)) = x(a\cos\theta + \sin\theta) + b\cos\theta .$$

Whenever $\cos\theta - a\sin\theta \neq 0$, in particular for θ sufficiently close to 0, the first equation is solvable for x, and

$$x = \frac{\tilde{x} + b\sin\theta}{\cos\theta - a\sin\theta} .$$

Therefore the transform of f by θ according to formula (7.3) is the linear function

$$\theta f(\tilde{x}) = \frac{\sin\theta + a\cos\theta}{\cos\theta - a\sin\theta} \cdot \tilde{x} + \frac{b}{\cos\theta - a\sin\theta} .$$

Definition 7.6. Given a system of partial differential equations, in p independent and q dependent variables, a symmetry group is a local group of transformations G acting on $X \times U$ such that whenever $u = f(x)$ is a local solution of this system, and for each g such that $g \cdot f$ is defined, then $\tilde{u} = gf(\tilde{x})$ is also a solution of the system.

For example, in the case of the heat equation $u_t = u_{xx}$, the group of translations in the spatial variable $(x, t, u) \mapsto (x+\lambda, t, u)$ is a symmetry group since $f(x+\lambda, t)$ is a solution of the heat equation whenever $f(x, t)$ is. Another example is provided by the group of Chapter 4 leaving the Navier-Stokes equations invariant. The ultimate goal of this chapter is to provide a readily verifiable criterion which will enable us to check whether or not a given group is a symmetry group of a given system of equations, and also find the symmetry group, meaning "the largest local group of symmetries," of the equations. The criterion will be infinitesimal, in direct analogy with the criterion of invariance of algebaraic equations of Theorem 7.3. In fact, when we finish constructing an appropriate machinery, we will be able to directly invoke Theorem 7.3 to prove invariance.

The key step is to construct spaces representing the various derivatives present in our system of partial differential equations and then concretely "realize" the system of equations as a subvariety of such a space. (This construction is a greatly simplified version of the construction of "jet bundles" in the differential-geometric theory of partial differential equations, cf [4] or [7].) Now for a given function of p independent variables, there are

$p_k = \binom{p+k-1}{k}$ different k-th order partial derivatives, which we denote by

$$\partial_J = \frac{\partial^{\Sigma J}}{\partial x_1^{j_1} \cdots \partial x_p^{j_p}} ,$$

where $J = (j_1, \ldots, j_p)$, each j_i is a nonnegative integer, and $\Sigma J = j_1 + \ldots + j_p = k$. Then given a function $f: X \to U$, so $u^\ell = f^\ell(x)$, there are $q \cdot p_k$ different numbers $u^\ell_J = \partial_J f^\ell(x)$ for $\ell = 1, \ldots, q$, $\Sigma J = k$ which give all the k-th order derivatives of f. Let $U_k = \mathbb{R}^{qp_k}$, with coordinates u^ℓ_J as above, be the space representing all these k-th order derivatives. Let $U^{(k)} = U \times U_1 \times \ldots \times U_k$ be the space representing all partial derivatives of functions $f: X \to U$ of order $\leq k$. Thus, given a function $f: X \to U$, there is a corresponding function $pr^{(k)} f: X \to U^{(k)}$, called the <u>k-th prolongation</u> of f, whose graph is given by the equations

$$u^\ell_J = \partial_J f^\ell(x) .$$

In other words, the value of $pr^{(k)} f(x)$ is a vector whose $q + qp_1 + \ldots + qp_k$ entries are the values of f and its partial derivatives of order $\leq k$ at the point x. Another way of looking at $pr^{(k)} f(x)$ is that it represents the Taylor polynomial of degree k of f at x, since the derivatives determine the Taylor polynomial and vice versa. (The total space $X \times U^{(k)}$ is called the k-jet space of $X \times U$, and the k-th prolongation of f is also called the k-jet of f in differential geometry.) We will use the symbol $u^{(k)}$ to denote points in $U^{(k)}$, so the entries of $u^{(k)}$ are the u_J^ℓ's.

A system of partial differential equations in p independent and q dependent variables is given by m equations of the form

$$\Delta^i(x, u^{(k)}) = 0 , \quad i = 1, \ldots, m.$$

Therefore, we may identify the system of equations with a subvariety $S_\Delta \subset X \times U^{(k)}$ given by the vanishing of a smooth function $\Delta: X \times U^{(k)} \to \mathbb{R}^m$, i.e.,

$$S_\Delta = \{(x, u^{(k)}): \Delta(x, u^{(k)}) = 0\} .$$

Then a <u>solution</u> of these equations is just a smooth function $f: X \to U$, such that $\Delta(x, pr^{(k)} f(x)) = 0$. In other words, the graph of the k-th prolongation of f lies entirely within S_Δ.

$$\{(x, pr^{(k)} f(x))\} \subset S_\Delta .$$

Now suppose G is a local group of transformations acting on the space $X \times U$. There is an induced local action of G on the space $X \times U^{(k)}$, called the <u>k-th prolongation</u> of G, and denoted by $pr^{(k)} G$. This prolonged action is such that the transform of the derivatives of a function is the derivatives of the transformed function, the latter being defined by (7.3). More rigorously, given $(x, u^{(k)}) \in X \times U^{(k)}$, choose a smooth function f defined in a neighborhood of x such that $\partial_J f^\ell(x) = u_J^\ell$. (Such a choice is always possible, for example f might be the finite Taylor polynomial at x corresponding to $u^{(k)}$:

$$f^\ell(\xi) = \sum_J \frac{u_J^\ell}{J!} (\xi - x)^J \text{ .)}$$

Then for $g \in G$ sufficiently close to the identity the function $g \cdot f$ is defined in a neighborhood of $(\tilde{x}, \tilde{u}) = g(x, u)$ by formula (7.3). We then define

$$pr^{(k)} g \cdot (x, u^{(k)}) = (\tilde{x}, \tilde{u}^{(k)}) \text{ .}$$

where the coordinates of $\tilde{u}^{(k)}$ are

$$\tilde{u}_J^\ell = \partial_J (g \cdot f)^\ell (\tilde{x}) \text{ .} \tag{7.4}$$

It is a straightforward matter to check that this definition is independent of the choice of function f to represent the point $(x, u^{(k)})$. The formula (7.4), when expanded using (7.3), will define the prolonged group action.

<u>Example 7.7.</u> Let U, X, G be as in example 7.5. Note first that $U_k \simeq \mathbb{R}$ with coordinate u_k representing the k-th derivative of a function $f(x)$. Thus

$$pr^{(k)}f(x) = (f(x), f'(x), \ldots, f^{(k)}(x)) .$$

We proceed to construct the first prolongation of the rotation group S_1. Let $(x^*, u^*, u_x^*) \in X \times U \times U_1 = X \times U^{(1)}$. Choose the linear Taylor polynomial

$$f(x) = u^* + (x - x^*)u_x^* = u^* - x^* u_x^* + xu^*$$

as a representative of (x^*, u^*, u_x^*), so that $f(x^*) = u^*$, $f'(x^*) = u_x^*$. According to the calculations of example 7.5, the transform of f by a rotation through an angle θ is the function

$$\theta \cdot f(\tilde{x}) = \frac{\sin\theta + u_x^* \cos\theta}{\cos\theta - u_x^* \sin\theta}\, \tilde{x} + \frac{u^* - x^* u_x^*}{\cos\theta - u_x^* \sin\theta} .$$

Now $\tilde{x}^* = x^* \cos\theta - u^* \sin\theta$, so

$$\tilde{u}^* = \theta \cdot f(\tilde{x}^*) = x^* \sin\theta + u^* \cos\theta ,$$

as we already knew. Moreover,

$$\tilde{u}_x^* = (\theta f)'(\tilde{x}^*) = \frac{\sin\theta + u_x^* \cos\theta}{\cos\theta - u_x^* \sin\theta} .$$

Therefore the prolonged action $pr^{(1)}S_1$ on $X \times U^{(1)}$ is given by

$$pr^{(1)}\theta \cdot (x, u, u_x) = \left(x\cos\theta - u\sin\theta, x\sin\theta + u\cos\theta, \frac{\sin\theta + u_x \cos\theta}{\cos\theta - u_x \sin\theta}\right),$$

which is defined whenever

$$|\theta| < |\operatorname{arc\,cot} u_x| .$$

Note that even though S_1 is a global transformation group, its prolongation $pr^{(1)}S_1$ is only a local transformation group. The infinitesimal generator of the prolonged group action is found by differentiating the last equation with respect to θ and setting $\theta = 0$. This yields

$$pr^{(1)}\alpha = -u\partial_x + x\partial_u + (1 + u_x^2)\partial_{u_x} .$$

(Here and in the sequel we will occasionally use ∂_x to denote $\partial/\partial x$, etc.)

Theorem 7.8. Suppose $\Delta = 0$ is a system of p.d.e.'s with corresponding subvariety $S_\Delta \subset X \times U^{(k)}$. Suppose S_Δ is invariant under $pr^{(k)}G$ for some group G acting on $X \times U$. Then G is a symmetry group of the system as in Definition 7.6.

Proof. Suppose $u = f(x)$ is a local solution. This means that the graph $\Gamma_f^{(k)}$, of $pr^{(k)}f$ lies entirely inside S_Δ. Now for $g \in G$ such that $g \cdot f$ is defined, the graph $\Gamma_{gf}^{(k)}$ of $pr^{(k)}(g\cdot f$ is just the transform of the graph of $pr^{(k)}f$ under $pr^{(k)}g$, i.e.,

$$\Gamma_{gf}^{(k)} = pr^{(k)}g \cdot \Gamma_f^{(k)} .$$

This is just a restatement of formula (7.4) defining $pr^{(k)}g$. Now since S_Δ is invariant under $pr^{(k)}g$, the graph of $pr^{(k)}(g \cdot f)$ lies entirely within S_Δ. But this is just another way of saying that $g \cdot f$ is a solution of $\Delta = 0$. This completes the proof.

Given an infinitesimal generator α of a one-parameter subgroup $\exp(t\alpha)$ of G, define the k-th prolongation of α to be the infinitesimal generator of the prolonged one-parameter subgroup $pr^{(k)}\exp(t\alpha)$; i.e.,

$$pr^{(k)}\alpha = \frac{d}{dt}\bigg|_{t=0} pr^{(k)}[\exp(t\alpha)]. \tag{7.5}$$

Combining Theorem 7.8 and Theorem 7.3, we get the following infinitesimal criterion for G to be the symmetry group of a system of p.d.e.'s.

Corollary 7.9. Suppose $\Delta(x, u^{(k)}) = 0$ is a system of p.d.e.'s (such that the Jacobian matrix of Δ has maximal rank everywhere). Suppose G is a connected local transformation group acting on $X \times U$ such that for every infinitesimal generator α of G

$$pr^{(k)}\alpha[\Delta(x, u^{(k)})] = 0 \tag{7.6}$$

whenever $\Delta(x, u^{(k)}) = 0$. Then G is a symmetry group of the equations $\Delta = 0$.

Example 7.10. Let X, U, G be as in examples 7.5, 7.7. Consider the first order ordinary differential equation

$$\Delta = (u-x)u_x + (u+x) = 0.$$

Applying the infinitesimal generator of $pr^{(1)}S_1$ to this equation yields

$$pr^{(1)}\alpha \cdot \Delta = (-u\partial_x + x\partial_u + (1+u_x^2)\partial_{u_x})[(u-x)u_x + (u+x)]$$

$$= u_x[(u-x)u_x + (u+x)]$$

$$= u_x \Delta .$$

Therefore $pr^{(1)}\alpha \cdot \Delta = 0$ whenever $\Delta = 0$, and condition (7.6) is verified. Then Corollary 7.10 shows that if $u = f(x)$ is any solution of $\Delta = 0$, then so is the rotated function $\tilde{u} = \theta \cdot f(\tilde{x})$. Indeed, in polar coordinates $x = r\cos\theta$, $u = r\sin\theta$, the equation $\Delta = 0$ becomes

$$dr/d\theta = r ,$$

whose solutions are the spirals

$$r = ce^{\theta} .$$

Obviously, any one of these spirals, when rotated, is another spiral of the same type. (For a discussion of the use of symmetry groups of ordinary differential equations for finding solutions by quadratures, the reader should consult reference [2].)

Theorem 7.8 and Corollary 7.9 admit converses if we further assume that the system of p. d. e.'s is "solvable for arbitrary initial dats." Then (7.6) becomes a necessary and sufficient condition for symmetry.

Theorem 7.11 *Suppose* $\Delta(x, u^{(k)}) = 0$ *is a system of partial differential equations in* p *independent variables* $x_1, \ldots, x_p$ *and* q *dependent variables* $u^1, \ldots, u^q$, *such that the Jacobian matrix of* Δ *has maximal rank everywhere. Suppose further that for any point*

$$(x_0, u_0^{(k)}) \in S_\Delta = \{(x, u^{(k)}) : \Delta(x, u^{(k)}) = 0\} \subset X \times U^{(k)}$$

there is a solution $u = f(x)$ *defined in a neighborhood of* x_0 *such that* $u_0^{(k)} = pr^{(k)} f(x_0)$. *Suppose* G *is a connected local transformation group acting on* $X \times U = \mathbb{R}^p \times \mathbb{R}^q$, *the space of independent and dependent variables. Then* G *is a symmetry group of the system if and only if for every infinitesimal generator* α *of* G,

$$pr^{(k)} \alpha[\Delta(x, u^{(k)})] = 0 \tag{7.7}$$

whenever $\Delta(x, u^{(k)}) = 0$.

Proof We need only show the necessity of (7.7). In view of Theorem 7.3, we must show that S_Δ is invariant under $pr^{(k)}G$, since this will imply (7.7). Given $(x_0, u_0^{(k)}) \in S_\Delta$, let $u = f(x)$ be a local solution with $u_0^{(k)} = pr^{(k)} f(x_0)$. For $g \in G$ such that $g \cdot f$ is defined,

$$pr^{(k)} g \cdot (x_0, u_0^{(k)}) = (\tilde{x}_0, pr^{(k)}(gf)(\tilde{x}_0)) \in S_\Delta ,$$

since gf is also a solution. This proves the theorem.

3. The Prolongation Formula

In light of Theorem 7.11, the primary task remaining is to find a formula for the prolongation of a vector field. Even though the prolonged group action, as determined by (7.4), is exceedingly complicated, we will find that the prolonged infinitesimal generators are expressed relatively simply. First we need the concept of a t total derivative.

Definition 7.12. Given a differentiable function $\Delta: X \times U^{(k)} \to \mathbb{R}$, the total derivative D_i $(1 \leq i \leq p)$ is the function $D_i\Delta: X \times U^{(k+1)} \to \mathbb{R}$ such that, for any smooth function $f: X \to U$,

$$D_i\Delta(x, pr^{(k+1)}f(x)) = \frac{\partial}{\partial x_i} \Delta(x, pr^{(k)}f(x)) .$$

In other words, $D_i\Delta$ is just the derivative of $\Delta(x, u^{(k)})$, treating u as a function of x.

It is easy to check that

$$D_i = \frac{\partial}{\partial x_i} + \sum_{\ell=1}^{q} \sum_{J} u^{\ell}_{J_i} \frac{\partial}{\partial u^{\ell}_{J}} , \tag{7.8}$$

where $J_i = (j_1, \ldots, j_{i-1}, j_i+1, j_{i+1}, \ldots, j_p)$, and the sum is over all J's with $j_1 + \ldots + j_p \leq k$. For instance, if $X = U = \mathbb{R}$, then there is just one total derivative

$$D_x = \frac{\partial}{\partial x} + u_x \frac{\partial}{\partial u} + u_{xx} \frac{\partial}{\partial u_x} + u_{xxx} \frac{\partial}{\partial u_{xx}} + \cdots .$$

Given a multi-index $J = (j_1, \ldots, j_p)$, we abbreviate

$$D^J = D_1^{j_1} D_2^{j_2} \cdots D_p^{j_p} .$$

Theorem 7.13. *Suppose* α *is a smooth vector field on* $X \times U$, *given by*

$$\alpha = \sum_{i=1}^{p} \xi^i(x,u) \frac{\partial}{\partial x_i} + \sum_{\ell=1}^{q} \varphi_\ell(x,u) \frac{\partial}{\partial u^\ell} .$$

The k-*th prolongation of* α , *as defined by* (7.5), *is the vector field*

$$\mathrm{pr}^{(k)}\alpha = \alpha + \sum_{\ell=1}^{q} \sum_{J} \varphi_\ell^J(x, u^{(k)}) \frac{\partial}{\partial u_J^\ell} \tag{7.9}$$

on $X \times U^{(k)}$, *where the sum is over all* J's *with* $0 < j_1 + \ldots + j_p \le k$. *The coefficient functions* φ_ℓ^J *are given by the following formula*:

$$\varphi_\ell^J = D^J(\varphi^\ell - \sum_{i=1}^{p} u_i^\ell \xi_i) + \sum_{i=1}^{p} u_{J_i}^\ell \xi^i , \tag{7.10}$$

where $u_i^\ell = \partial u^\ell / \partial x_i$ *and* J_i *is as defined above.*

Proof. First the formula will be proved for the case $k = 1$.

Let

$$(\tilde{x}_t, \tilde{u}_t) = \exp(t\alpha)(x,u) = (\Xi_t(x,u), \Phi_t(x,u)) ,$$

so that

$$\left.\frac{d}{dt}\right|_{t=0} \Xi_t^i(x,u) = \xi^i(x,u) \qquad i = 1,\dots,p,$$

$$\left.\frac{d}{dt}\right|_{t=0} \Phi_t^\ell(x,u) = \varphi_\ell(x,u) \qquad \ell = 1,\dots,q.$$

Now given $(x, u^{(1)}) \in X \times U^{(1)}$, let $u = f(x)$ be any representative, so that $u_i^\ell = \partial f(x)/\partial x_i$. According to (7.3), for t sufficiently small, the transform of f by the group element $\exp(t\alpha)$ is well-defined and is given by

$$f_t(\tilde{x}_t) = [\Phi_t \circ (I \times f)] \circ [\Xi_t \circ (I \times f)]^{-1}(x_t) .$$

Using the chain rule, the Jacobian matrix of f_t at x_t is therefore

$$Jf_t(\tilde{x}_t) = J[\Phi_t \circ (I \times f)](x) \cdot [J[\Xi_t \circ (I \times f)](x)]^{-1} . \tag{7.11}$$

This serves to define the prolonged group action $\mathrm{pr}^{(1)}\exp(t\alpha)$. Thus to find the infinitesimal generator $\mathrm{pr}^{(1)}\alpha$, we must differentiate (7.11) with respect to t and set $t = 0$. Recall that for any matrix valued function $A(t)$,

$$\frac{d}{dt}[A^{-1}(t)] = -A^{-1}(t)\,\frac{dA(t)}{dt}\,A^{-1}(t) .$$

Also note that since $t = 0$ corresponds to the identity group element,

$$\Xi_0 \circ (I \times f) = I , \qquad \Phi_0 \circ (I \times f) = f .$$

Therefore, by Leibnitz' rule,

$$\frac{d}{dt}\bigg|_{t=0} Jf_t(\tilde{x}_t) = \frac{d}{dt}\bigg|_{t=0} J[\Phi_t \circ (I \times f)](x) -$$

$$- Jf(x) \cdot \frac{d}{dt}\bigg|_{t=0} J[\Xi_t \circ (I \times f)](x)$$

$$= J[\varphi \circ (I \times f)](x) - Jf(x) \cdot J[\xi \circ (I \times f)](x) .$$

Now the matrix entries of this are just the coordinate functions of the first prolongation of α ; namely

$$\varphi^j_\ell = \frac{\partial}{\partial x_j}[\varphi_\ell(x, f(x))] - \sum_{i=1}^{p} \frac{\partial f^\ell}{\partial x_i} \cdot \frac{\partial}{\partial x_j}[\xi^i(x, f(x))]$$

$$= D_j \varphi_\ell (x, u^{(1)}) - \sum_{i=1}^{p} u^\ell_i D_j \xi^i(x, u^{(1)})$$

$$= D_j[\varphi^\ell - \sum_{i=1}^{p} u^\ell_i \xi^i] \quad \sum_{i=1}^{p} u^\ell_{ij} \xi^i ,$$

where we have used the definition of the total derivative, and $u^\ell_{ij} = \partial^2 u^\ell / \partial x_i \partial x_j$. This proves the theorem when $k = 1$.

To prove the theorem in general, we proceed by induction. Notice that $X \times U^{(k+1)}$ can be viewed as a subspace of $X \times [U^{(k)}]^{(1)}$. Therefore given a multi-index J, by what we have already proven,

$$\varphi^{J_j}_\ell = D_j \varphi^J_\ell - \sum_{i=1}^{p} u^\ell_{J_i} D_j \xi^i . \qquad (7.12)$$

(Equation (7.10) is a useful recursion relation for the φ_ℓ^J's, and may be also found in [3; page 106].) It is a simple matter to check that (7.10) satisfies the recursion relation (7.12). Indeed,

$$\begin{aligned} D_j\varphi_\ell^J - \sum_i u^\ell_{J_i} D_j\xi^i &= D^{J_j}(\varphi^\ell - \sum_i u_i^\ell \xi^i) + \\ &+ \sum_i (u^\ell_{J_{ij}} \xi^i + u^\ell_{J_i} D_j\xi^i) - \sum_i u^\ell_{J_i} D_j\xi^i \\ &= D^{J_j}(\varphi^\ell - \sum_i u_i^\ell \xi^i) + \sum_i u^\ell_{J_{ij}} \xi^i . \end{aligned}$$

(Here $J_{ij} = (J_i)_j$.) This completes the proof of Theorem 7.13.

<u>Example 7.14.</u> Let X, U, G be as in examples 7.5, 7.7 and 7.10. The infinitesimal generator of the rotation group is $\alpha = -u\partial_x + x\partial_u$. The first prolongation, according to (7.9, 10), is the vector field $pr^{(1)}\alpha = \alpha + \varphi^x \partial_{u_x}$, where

$$\begin{aligned} \varphi^x &= D_x(x + uu_x) - uu_{xx} \\ &= 1 + u_x^2 , \end{aligned}$$

as we have already discovered. Similarly, the coefficient function φ^{xx} of $\partial/\partial u_{xx}$ in the second prolongation of α is

$$\varphi^{xx} = D_x^2(x + u_x^2 + uu_{xx}) - uu_{xxx}$$

$$= D_x(1 + u_x^2 + uu_{xx}) - uu_{xxx}$$

$$= 3u_x u_{xx} .$$

Therefore the infinitesimal generator of the second prolongation of the rotation group is the vector field

$$pr^{(2)}\alpha = -u\partial_x + x\partial_u + (1 + u_x^2)\partial_{u_x} + 3u_x u_{xx}\partial_{u_{xx}} .$$

(The reader is invited to attempt to deduce this formula directly from the prolonged group action!) Using the infinitesimal criterion of Theorem 7.11, we see that the differential equation $u_{xx} = 0$ is invariant under S_1, since $pr^{(2)}\alpha(u_{xx}) = 3u_x u_{xx} = 0$ whenever $u_{xx} = 0$. This is just a restatement of the fact that rotations preserve straight lines. Similarly the function $F(u_x, u_{xx}) = u_{xx}(1+u_x^2)^{-3/2}$ is invariant under $pr^{(2)}S_1$ since $pr^{(2)}\alpha \cdot F = 0$. This just says that the curvature of a curve is invariant under rotations.

Example 7.15. Consider the case $p = 2$, $q = 1$, so we are looking at partial differential equations for functions $u = f(x,t)$. A vector field on $X \times U$ is of the form

$$\alpha = \xi\partial_x + \tau\partial_t + \varphi\partial_u ,$$

where ξ, τ, φ are functions of x, t, u. The first prolongation of α is the vector field

$$\mathrm{pr}^{(1)}\alpha = \alpha + \varphi^{x}\partial_{u_x} + \varphi^{t}\partial_{u_t} ,$$

where

$$\begin{aligned}
\varphi^{x} &= D_x(\varphi - u_x\xi - u_t\tau) + u_{xx}\xi + u_{xt}\tau \\
&= D_x\varphi - u_x D_x\xi - u_t D_x\tau \\
&= \varphi_x + (\varphi_u - \xi_x)u_x - \tau_x u_t - \xi_u u_x^2 - \tau_u u_x u_t \\
\varphi^{t} &= D_t(\varphi - u_x\xi - u_t\tau) + u_{xt}\xi + u_{tt}\tau \\
&= D_t\varphi - u_x D_t\xi - u_t D_t\tau \\
&= \varphi_t - \xi_t u_x + (\varphi_u - \tau_t)u_t - \xi_u u_x u_t - \tau_u u_x^2 .
\end{aligned} \tag{7.13}$$

Similarly, the second prolongation of α is

$$\mathrm{pr}^{(2)}\alpha = \mathrm{pr}^{(1)}\alpha + \varphi^{xx}\partial_{u_{xx}} + \varphi^{xt}\partial_{u_{xt}} + \varphi^{tt}\partial_{u_{tt}}$$

where, for example,

$$\varphi^{xx} = D_x^2(\varphi - u_x\xi - u_t\tau) + u_{xxx}\xi + u_{xxt}\tau$$

$$= D_x^2\varphi - u_x D_x^2\xi - u_t D_x^2\tau - 2u_{xx}D_x\xi - 2u_{xt}D_x\tau$$

$$= \varphi_x + u_x(2\varphi_{xu} - \xi_{xx}) - u_t\tau_{xx} + \tag{7.14}$$

$$+ u_x^2(\varphi_{uu} - 2\xi_{xu}) - u_x u_t \tau_{xu} - u_x^3\xi_{uu} - u_x^2 u_t \tau_{uu} +$$

$$+ u_{xx}(\varphi_u - 2\xi_x) - 2u_{xt}\tau_x - 3u_{xx}u_x\xi_u - u_{xx}u_t\tau_u - 2u_{xt}u_x\tau_u .$$

These expressions will be used in the following section to compute the symmetry group of the heat equation.

Theorem 7.16. Suppose α and β are smooth vector fields on $X \times U$. Then

$$pr^{(k)}[\alpha, \beta] = [pr^{(k)}\alpha, pr^{(k)}\beta] .$$

Corollary 7.17. If $\Delta(x, u^{(k)}) = 0$ is a system of p.d.e.'s satisfying the conditions of Theorem 7.12, then the set of all infinitesimal symmetries of $\Delta = 0$, meaning the set of all vector fields α on $X \times U$ generating one-parameter symmetry groups, is a Lie algebra.

The most straightforward proof of 7.16 is computational using the prolongation formula of Theorem 7.13. The details are left to the reader. (See also [4].)

4. Applications of the Theory.

The basic method for finding the (connected component) of the symmetry group of a given system of p.d.e.'s is to substitute the prolongation formula of Theorem 7.13 for a vector field α on $X \times U$ into the infinitesimal criterion of invariance (7.6). The coefficients of the various partial derivatives of the dependent variables in the resulting equations are equated, which gives a large system of elementary p.d.e.'s for the coefficient functions of α, called the symmetry equations. The general solution of the symmetry equations is then the most general infinitesimal symmetry of the given system. The symmetry group itself may be found via exponentiation.

As a first example, consider the one-dimensional heat equation

$$u_t = u_{xx} \quad . \tag{7.15}$$

Note that $p = 2$, $q = 1$ and $k = 2$; the heat equation being the linear subvariety of $X \times U^{(2)}$ given by the vanishing of the function

$\Delta = u_t - u_{xx}$. Given a vector field $\alpha = \xi\partial_x + \tau\partial_t + \varphi\partial_u$ on $X \times U$, the second prolongation $pr^{(2)}\alpha$ is given in example 7.15. The infinitesimal criterion (7.6) is just

$$\varphi^t = \varphi^{xx} , \tag{7.16}$$

which must be satisfied whenever $u_t = u_{xx}$. Substituting (7.13, 14) into (7.16), replacing u_t by u_{xx}, and equating the coefficients of the various partial derivatives of u, yields the following system of equations:

$u_{xx}u_{xt}$:	$0 = -\tau_u$	(a)
u_{xt} :	$0 = -2\tau_x$	(b)
u_{xx}^2 :	$-\tau_u = -\tau_u$	(c)
$u_x^2 u_{xx}$:	$0 = -\tau_{uu}$	(d)
$u_x u_{xx}$:	$-\xi_u = -2\tau_{xu} - 3\xi_u$	(e)
u_{xx} :	$\varphi_u - \tau_t = -\tau_{xx} + \varphi_u - 2\xi_x$	(f)
u_x^3:	$0 = -\xi_{uu}$	(g)
u_x^2:	$0 = \varphi_{uu} - 2\xi_{xu}$	(h)
u_x:	$-\xi_t = 2\varphi_{xu} - \xi_{xx}$	(i)
1:	$\varphi_t = \varphi_{xx}$	(j)

These are the symmetry equations. Now (a) and (b) show that τ is just a function of t. Then (e) shows that ξ depends only on x, t, and (f) shows $\tau_t = 2\xi_x$, hence $\xi_{xx} = 0$. Then (h) shows that $\varphi = g(x,t)u + f(x,t)$, and by (i), $\xi_t = -2g_x$. Finally (j) implies $g_t = g_{xx}$ and $f_t = f_{xx}$, hence $g_{xxx} = 0 = g_{xt}$, so $\xi_{tt} = 0$. Therefore the most general solution of the symmetry equations is

$$\begin{aligned} \xi &= c_1 + c_4 x + 2c_5 t + 4c_6 xt , \\ \tau &= c_2 + 2c_4 t + 4c_6 t^2 , \\ \varphi &= (c_3 - c_5 x - 2c_6 t - c_6 x^2)u + f(x,t) , \end{aligned} \tag{7.17}$$

where $c_1, \ldots, c_6$ are arbitrary constants and $f(x,t)$ an arbitrary solution of the heat equation. Thus the Lie algebra of infinitesimal symmetries of the heat equation is spanned by the six vector fields

$$\begin{aligned} \alpha_1 &= \partial_x \\ \alpha_2 &= \partial_t \\ \alpha_3 &= u\partial_u \\ \alpha_4 &= x\partial_x + 2t\partial_t \\ \alpha_5 &= 2t\partial_x - xu\partial_u \\ \alpha_6 &= 4tx\partial_x + 4t^2\partial_t - (x^2 + 2t)u\partial_u \end{aligned} \tag{7.18}$$

and the infinite-dimensional subalgebra

$$\alpha_f = f(x,t)\partial_u$$

where f is an arbitrary solution of the heat equation. The one-parameter groups $G_i = \exp(\lambda\alpha_i)$ generated by α_i are given by the expressions

$$
\begin{array}{ll}
G_1: & (x + \lambda, t, u) \\
G_2: & (x, t + \lambda, u) \\
G_3: & (x, t, e^{\lambda}u) \qquad \lambda \in \mathbb{R} \\
G_4: & (e^{\lambda}x, e^{2\lambda}t, u) \\
G_5: & (x - 2\lambda t, t, u\exp(x\lambda - \lambda^2 t)) \\
G_6: & \left(\frac{x}{4\lambda t + 1}, \frac{t}{4\lambda t + 1}, u\sqrt{4\lambda t + 1}\ \exp[\frac{-\lambda x^2}{4\lambda t + 1}]\right) \\
G_f: & (x, t, u + \lambda f(x, t)).
\end{array}
\tag{7.19}
$$

Note that the symmetries G_3, G_f are consequences of the fact that the heat equation is linear. G_1 and G_2 reflect the fact that the heat equation has constant coefficients. G_4 is the well-known scale symmetry, G_5 a kind of Gallilean boost. This result is not new, see for instance [1] and [2] for a more complete discussion of these symmetries. Note also that G_6 is a local group.

For our second example, the symmetry group of the Navier-Stokes equations

$$
\begin{aligned}
p_x + uu_x + vu_y + wu_z &= \Delta u \\
p_y + uv_x + vv_y + wv_z &= \Delta v \\
p_z + uw_x + vw_y + ww_z &= \Delta w \\
u_x + v_y + w_z &= 0
\end{aligned}
\tag{7.20}
$$

will be computed. In this case $p = 3$, with coordinates (x, y, z), and $q = 4$, with coordinates (u, v, w, p). A typical vector field on $X \times U \simeq \mathbb{R}^3 \times \mathbb{R}^4$ is given by

$$\alpha = \xi\partial_x + \eta\partial_y + \zeta\partial_z + \varphi\partial_u + \psi\partial_v + \chi\partial_w + \pi\partial_p ,$$

where the coefficient functions depend on (x, y, z, u, v, w, p). The infinitesimal symmetry criterion (7.6) for (7.20) is

$$\begin{aligned}
\pi^x + u\varphi^x + v\varphi^y + w\varphi^z + u_x\varphi + u_y\psi + u_z\chi &= \varphi^{xx} + \varphi^{yy} + \varphi^{zz} \\
\pi^y + u\psi^x + v\psi^y + w\psi^z + v_x\varphi + v_y\psi + v_z\chi &= \psi^{xx} + \psi^{yy} + \psi^{zz} , \qquad (7.21) \\
\pi^z + u\chi^x + v\chi^y + w\chi^z + w_x\varphi + w_y\psi + w_z\chi &= \chi^{xx} + \chi^{yy} + \chi^{zz} , \\
\varphi^x + \psi^y + \chi^z &= 0 ,
\end{aligned}$$

which must be satisfied whenever (7.20) is. Here φ^x, etc. are the coefficient functions of the second prolongation of α, and are given by the prolongation formula (7.10). (See also (7.13, 14) for prototypical examples.) In the first equation of (7.21), the coefficient of p_{xx} is

$$0 = \varphi_p - u_x\xi_p - u_y\eta_p - u_z\zeta_p ,$$

hence, $\xi, \eta, \zeta, \varphi$, and, by similar arguments, ψ and χ do not depend on p. Next the coefficient of u_{xy} and other mixed second-order derivatives of u, v and w in the first three equations of (7.21) shows that

$$\xi_y + \eta_x = 0, \qquad \xi_z + \zeta_x = 0, \qquad \eta_z + \zeta_y = 0 \tag{7.22}$$

If, in the first equation of (7.21), we replace p_x by its value as given in (7.20), then the resulting coefficient of v_{yy} shows that $\varphi_v + \eta_x = 0$. Similarly, we find

$$\begin{aligned} &\varphi_v + \eta_x = 0 \qquad && \psi_u + \xi_y = 0 \qquad && \chi_u + \xi_z = 0, \\ &\varphi_w + \zeta_x = 0 \qquad && \psi_w + \zeta_y = 0 \qquad && \chi_v + \eta_z = 0. \end{aligned} \tag{7.23}$$

Next the coefficient of u_{xx} (again after replacing p_x) in the first equation of (7.21) shows that ξ (and, similarly, η and ζ) depends only on x, y, z. Moreover, we find

$$\begin{aligned} \varphi_u &= \pi_p + \xi_x, \\ \psi_v &= \pi_p + \eta_y, \\ \chi_w &= \pi_p + \zeta_z \end{aligned} \tag{7.24}$$

Note that this implies that π depends linearly on p. The coefficient of u_y in the first equation of (7.21), using (7.24), yields

$$\psi = \eta_x u + (\eta_y - 2\xi_x)v + \eta_z w - \Delta\eta + 2\varphi_{yu} .$$

Comparison with (7.24) requires that $\pi_p = -2\xi_x$. Therefore we get the following representations of φ, ψ, and χ :

$$\varphi = -\xi_x u + \xi_y v + \xi_z w - 3\xi_{xx} - \xi_{yy} - \xi_{xx} ,$$

$$\psi = \eta_x u - \eta_y v + \eta_z w - \eta_{xx} - 3\eta_{yy} - \eta_{zz} , \tag{7.25}$$

$$\chi = \zeta_x u + \zeta_y v - \zeta_z w - \zeta_{xx} - \zeta_{yy} - 3\zeta_{zz} .$$

Moreover

$$\xi_x = \eta_y = \zeta_z = 0 . \tag{7.26}$$

Consideration of other first derivatives of u, v, w in the first three equations of (7.20) shows that π is linear in u, v, w and p. Also the mixed partial derivatives of ξ, η, ζ, for instance, ξ_{xz}, are all 0. Finally the term in the first equation of (7.20) not involving any derivatives of u, v, w, p is

$$\pi_x + u\varphi_x + v\varphi_t + 2\varphi_z = \Delta\varphi .$$

The quadratic terms u^2, v^2, and w^2 of this last equation imply that $\xi_{xx} = \xi_{yy} = \xi_{zz} = 0$, so that ξ, and also η and ζ, must be linear in x, y, z. The general solution of (7.21) is then found to be given by

$$\begin{aligned}
\xi &= c_1 + c_7 x + c_4 y + c_5 z \\
\eta &= c_2 - c_4 x + c_7 y + c_6 z \\
\zeta &= c_3 - c_5 x - c_6 y + c_7 z \\
\varphi &= -c_7 u + c_4 v + c_3 w \\
\psi &= -c_4 u - c_7 v + c_6 w \\
\chi &= -c_5 u - c_6 v - c_7 w \\
\pi &= c_8 - 2c_7 p .
\end{aligned}$$

Therefore the Lie algebra of infinitesimal symmetries of the Navier-Stokes equations is spanned by the vector fields

$$\alpha_1 = \partial_x$$

$$\alpha_2 = \partial_y$$

$$\alpha_3 = \partial_z$$

$$\alpha_4 = y\partial_x - x\partial_y + v\partial_u - u\partial_v$$

$$\alpha_5 = z\partial_x - x\partial_z + w\partial_u - u\partial_w$$

$$\alpha_6 = z\partial_y - y\partial_z + w\partial_v - v\partial_w$$

$$\alpha_7 = x\partial_x + y\partial_y + z\partial_z - u\partial_u - v\partial_v - w\partial_w - 2p\partial_p$$

$$\alpha_8 = \partial_p \ .$$

It is easily verified that the first six of these vector fields are just the infinitesimal generators of the action of the group of rigid motions given previously in Chapter 4, section 5. Hence $\alpha_1, \alpha_2, \alpha_3$ generate the translations, and $\alpha_4, \alpha_5, \alpha_6$ the rotations. The vector field α_7 generates a group of scale transformations:

$$G_7: \quad (e^{\lambda}x, \ e^{\lambda}y, \ e^{\lambda}z, \ e^{-\lambda}u, \ e^{-\lambda}v, \ e^{-\lambda}w, \ e^{-2\lambda}p).$$

This means that if

$$(u, v, w, p) = (f(x,y,z), \ g(x,y,z), \ h(x,y,z), j(x,y,z))$$

is a solution of the Navier-Stokes equations, so is

$$(\tilde{u}, \tilde{v}, \tilde{w}, \tilde{p}) = (e^{-\lambda}f(e^{-\lambda}x, e^{-\lambda}y, e^{-\lambda}z), e^{-\lambda}g, e^{-\lambda}h, e^{-2\lambda}j),$$

where g, h, j are also evaluated at $e^{-\lambda}(x,y,z)$. Finally, α_φ comes from the fact that the Navier-Stokes equations are invariant under pressure translations.

References

1. Bluman, G. W. and Cole, J. D., "The General Similarity Solution of the Heat Equation," J. Math. Mech., (11) 18 (1969), pp. 1025-1042.

2. Bluman, G. W. and Cole, J. D., Similarity Methods for Differential Equations, Springer-Verlag, Applied Math. Sci. No. 13, New York, 1974.

3. Eisenhart, L. P., Continuous Groups of Transformations, Princeton University Press, Princeton, N. J., 1933.

4. Olver, P. J., "Symmetry Groups and Group Invariant Solutions of Partial Differential Equations," to appear, J. Diff. Geom.

5. Ovsjannikov, L. V., Group Properties of Differential Equations, transl. by G. W. Bluman, 1967 (unpublished).

6. Palais, R. S., "A Global Formulation of the Lie Theory of Transformation Groups," Memoirs of the A. M. S. No. 22, Providence, R. J., 1957.

7. Palais, R. S., ed., Seminar on the Atiyah-Singer Index Theorem, Annals of Math Studies, No. 57, Princeton University Press, Princeton, N. J., 1965. (Chapter 4).

8. Warner, F. W., Foundations of Differentiable Manifolds and Lie Groups, Scott, Foresman and Company, Glenview, Ill. 1971.

Subject Index

Vol. 609: General Topology and Its Relations to Modern Analysis and Algebra IV. Proceedings 1976. Edited by J. Novák. XVIII, 225 pages. 1977.

Vol. 610: G. Jensen, Higher Order Contact of Submanifolds of Homogeneous Spaces. XII, 154 pages. 1977.

Vol. 611: M. Makkai and G. E. Reyes, First Order Categorical Logic. VIII, 301 pages. 1977.

Vol. 612: E. M. Kleinberg, Infinitary Combinatorics and the Axiom of Determinateness. VIII, 150 pages. 1977.

Vol. 613: E. Behrends et al., L^p-Structure in Real Banach Spaces. X, 108 pages. 1977.

Vol. 614: H. Yanagihara, Theory of Hopf Algebras Attached to Group Schemes. VIII, 308 pages. 1977.

Vol. 615: Turbulence Seminar, Proceedings 1976/77. Edited by P. Bernard and T. Ratiu. VI, 155 pages. 1977.

Vol. 616: Abelian Group Theory, 2nd New Mexico State University Conference, 1976. Proceedings. Edited by D. Arnold, R. Hunter and E. Walker. X, 423 pages. 1977.

Vol. 617: K. J. Devlin, The Axiom of Constructibility: A Guide for the Mathematician. VIII, 96 pages. 1977.

Vol. 618: I. I. Hirschman, Jr. and D. E. Hughes, Extreme Eigen Values of Toeplitz Operators. VI, 145 pages. 1977.

Vol. 619: Set Theory and Hierarchy Theory V, Bierutowice 1976. Edited by A. Lachlan, M. Srebrny, and A. Zarach. VIII, 358 pages. 1977.

Vol. 620: H. Popp, Moduli Theory and Classification Theory of Algebraic Varieties. VIII, 189 pages. 1977.

Vol. 621: Kauffman et al., The Deficiency Index Problem. VI, 112 pages. 1977.

Vol. 622: Combinatorial Mathematics V, Melbourne 1976. Proceedings. Edited by C. Little. VIII, 213 pages. 1977.

Vol. 623: I. Erdelyi and R. Lange, Spectral Decompositions on Banach Spaces. VIII, 122 pages. 1977.

Vol. 624: Y. Guivarc'h et al., Marches Aléatoires sur les Groupes de Lie. VIII, 292 pages. 1977.

Vol. 625: J. P. Alexander et al., Odd Order Group Actions and Witt Classification of Innerproducts. IV, 202 pages. 1977.

Vol. 626: Number Theory Day, New York 1976. Proceedings. Edited by M. B. Nathanson. VI, 241 pages. 1977.

Vol. 627: Modular Functions of One Variable VI, Bonn 1976. Proceedings. Edited by J.-P. Serre and D. B. Zagier. VI, 339 pages. 1977.

Vol. 628: H. J. Baues, Obstruction Theory on the Homotopy Classification of Maps. XII, 387 pages. 1977.

Vol. 629: W.A. Coppel, Dichotomies in Stability Theory. VI, 98 pages. 1978.

Vol. 630: Numerical Analysis, Proceedings, Biennial Conference, Dundee 1977. Edited by G. A. Watson. XII, 199 pages. 1978.

Vol. 631: Numerical Treatment of Differential Equations. Proceedings 1976. Edited by R. Bulirsch, R. D. Grigorieff, and J. Schröder. X, 219 pages. 1978.

Vol. 632: J.-F. Boutot, Schéma de Picard Local. X, 165 pages. 1978.

Vol. 633: N. R. Coleff and M. E. Herrera, Les Courants Résiduels Associés à une Forme Méromorphe. X, 211 pages. 1978.

Vol. 634: H. Kurke et al., Die Approximationseigenschaft lokaler Ringe. IV, 204 Seiten. 1978.

Vol. 635: T. Y. Lam, Serre's Conjecture. XVI, 227 pages. 1978.

Vol. 636: Journées de Statistique des Processus Stochastiques, Grenoble 1977, Proceedings. Edité par Didier Dacunha-Castelle et Bernard Van Cutsem. VII, 202 pages. 1978.

Vol. 637: W. B. Jurkat, Meromorphe Differentialgleichungen. VII, 194 Seiten. 1978.

Vol. 638: P. Shanahan, The Atiyah-Singer Index Theorem, An Introduction. V, 224 pages. 1978.

Vol. 639: N. Adasch et al., Topological Vector Spaces. V, 125 pages. 1978.

Vol. 640: J. L. Dupont, Curvature and Characteristic Classes. X, 175 pages. 1978.

Vol. 641: Séminaire d'Algèbre Paul Dubreil, Proceedings Paris 1976–1977. Edité par M. P. Malliavin. IV, 367 pages. 1978.

Vol. 642: Theory and Applications of Graphs, Proceedings, Michigan 1976. Edited by Y. Alavi and D. R. Lick. XIV, 635 pages. 1978.

Vol. 643: M. Davis, Multiaxial Actions on Manifolds. VI, 141 pages. 1978.

Vol. 644: Vector Space Measures and Applications I, Proceedings 1977. Edited by R. M. Aron and S. Dineen. VIII, 451 pages. 1978.

Vol. 645: Vector Space Measures and Applications II, Proceedings 1977. Edited by R. M. Aron and S. Dineen. VIII, 218 pages. 1978.

Vol. 646: O. Tammi, Extremum Problems for Bounded Univalent Functions. VIII, 313 pages. 1978.

Vol. 647: L. J. Ratliff, Jr., Chain Conjectures in Ring Theory. VIII, 133 pages. 1978.

Vol. 648: Nonlinear Partial Differential Equations and Applications, Proceedings, Indiana 1976–1977. Edited by J. M. Chadam. VI, 206 pages. 1978.

Vol. 649: Séminaire de Probabilités XII, Proceedings, Strasbourg, 1976–1977. Edité par C. Dellacherie, P. A. Meyer et M. Weil. VIII, 805 pages. 1978.

Vol. 650: C*-Algebras and Applications to Physics. Proceedings 1977. Edited by H. Araki and R. V. Kadison. V, 192 pages. 1978.

Vol. 651: P. W. Michor, Functors and Categories of Banach Spaces. VI, 99 pages. 1978.

Vol. 652: Differential Topology, Foliations and Gelfand-Fuks-Cohomology, Proceedings 1976. Edited by P. A. Schweitzer. XIV, 252 pages. 1978.

Vol. 653: Locally Interacting Systems and Their Application in Biology. Proceedings, 1976. Edited by R. L. Dobrushin, V. I. Kryukov and A. L. Toom. XI, 202 pages. 1978.

Vol. 654: J. P. Buhler, Icosahedral Golois Representations. III, 143 pages. 1978.

Vol. 655: R. Baeza, Quadratic Forms Over Semilocal Rings. VI, 199 pages. 1978.

Vol. 656: Probability Theory on Vector Spaces. Proceedings, 1977. Edited by A. Weron. VIII, 274 pages. 1978.

Vol. 657: Geometric Applications of Homotopy Theory I, Proceedings 1977. Edited by M. G. Barratt and M. E. Mahowald. VIII, 459 pages. 1978.

Vol. 658: Geometric Applications of Homotopy Theory II, Proceedings 1977. Edited by M. G. Barratt and M. E. Mahowald. VIII, 487 pages. 1978.

Vol. 659: Bruckner, Differentiation of Real Functions. X, 247 pages. 1978.

Vol. 660: Equations aux Dérivée Partielles. Proceedings, 1977. Edité par Pham The Lai. VI, 216 pages. 1978.

Vol. 661: P. T. Johnstone, R. Paré, R. D. Rosebrugh, D. Schumacher, R. J. Wood, and G. C. Wraith, Indexed Categories and Their Applications. VII, 260 pages. 1978.

Vol. 662: Akin, The Metric Theory of Banach Manifolds. XIX, 306 pages. 1978.

Vol. 663: J. F. Berglund, H. D. Junghenn, P. Milnes, Compact Right Topological Semigroups and Generalizations of Almost Periodicity. X, 243 pages. 1978.

Vol. 664: Algebraic and Geometric Topology, Proceedings, 1977. Edited by K. C. Millett. XI, 240 pages. 1978.

Vol. 665: Journées d'Analyse Non Linéaire. Proceedings, 1977. Edité par P. Bénilan et J. Robert. VIII, 256 pages. 1978.

Vol. 666: B. Beauzamy, Espaces d'Interpolation Réels: Topologie et Géometrie. X, 104 pages. 1978.

Vol. 667: J. Gilewicz, Approximants de Padé. XIV, 511 pages. 1978.

Vol. 668: The Structure of Attractors in Dynamical Systems. Proceedings, 1977. Edited by J. C. Martin, N. G. Markley and W. Perrizo. VI, 264 pages. 1978.

Vol. 669: Higher Set Theory. Proceedings, 1977. Edited by G. H. Müller and D. S. Scott. XII, 476 pages. 1978.

Vol. 670: Fonctions de Plusieurs Variables Complexes III, Proceedings, 1977. Edité par F. Norguet. XII, 394 pages. 1978.

Vol. 671: R. T. Smythe and J. C. Wierman, First-Passage Perculation on the Square Lattice. VIII, 196 pages. 1978.

Vol. 672: R. L. Taylor, Stochastic Convergence of Weighted Sums of Random Elements in Linear Spaces. VII, 216 pages. 1978.

Vol. 673: Algebraic Topology, Proceedings 1977. Edited by P. Hoffman, R. Piccinini and D. Sjerve. VI, 278 pages. 1978.

Vol. 674: Z. Fiedorowicz and S. Priddy, Homology of Classical Groups Over Finite Fields and Their Associated Infinite Loop Spaces. VI, 434 pages. 1978.

Vol. 675: J. Galambos and S. Kotz, Characterizations of Probability Distributions. VIII, 169 pages. 1978.

Vol. 676: Differential Geometrical Methods in Mathematical Physics II, Proceedings, 1977. Edited by K. Bleuler, H. R. Petry and A. Reetz. VI, 626 pages. 1978.

Vol. 677: Séminaire Bourbaki, vol. 1976/77, Exposés 489–506. IV, 264 pages. 1978.

Vol. 678: D. Dacunha-Castelle, H. Heyer et B. Roynette. Ecole d'Eté de Probabilités de Saint-Flour. VII-1977. Edité par P. L. Hennequin. IX, 379 pages. 1978.

Vol. 679: Numerical Treatment of Differential Equations in Applications, Proceedings, 1977. Edited by R. Ansorge and W. Törnig. IX, 163 pages. 1978.

Vol. 680: Mathematical Control Theory, Proceedings, 1977. Edited by W. A. Coppel. IX, 257 pages. 1978.

Vol. 681: Séminaire de Théorie du Potentiel Paris, No. 3, Directeurs: M. Brelot, G. Choquet et J. Deny. Rédacteurs: F. Hirsch et G. Mokobodzki. VII, 294 pages. 1978.

Vol. 682: G. D. James, The Representation Theory of the Symmetric Groups. V, 156 pages. 1978.

Vol. 683: Variétés Analytiques Compactes, Proceedings, 1977. Edité par Y. Hervier et A. Hirschowitz. V, 248 pages. 1978.

Vol. 684: E. E. Rosinger, Distributions and Nonlinear Partial Differential Equations. XI, 146 pages. 1978.

Vol. 685: Knot Theory, Proceedings, 1977. Edited by J. C. Hausmann. VII, 311 pages. 1978.

Vol. 686: Combinatorial Mathematics, Proceedings, 1977. Edited by D. A. Holton and J. Seberry. IX, 353 pages. 1978.

Vol. 687: Algebraic Geometry, Proceedings, 1977. Edited by L. D. Olson. V, 244 pages. 1978.

Vol. 688: J. Dydak and J. Segal, Shape Theory. VI, 150 pages. 1978.

Vol. 689: Cabal Seminar 76–77, Proceedings, 1976–77. Edited by A.S. Kechris and Y. N. Moschovakis. V, 282 pages. 1978.

Vol. 690: W. J. J. Rey, Robust Statistical Methods. VI, 128 pages. 1978.

Vol. 691: G. Viennot, Algèbres de Lie Libres et Monoïdes Libres. III, 124 pages. 1978.

Vol. 692: T. Husain and S. M. Khaleelulla, Barrelledness in Topological and Ordered Vector Spaces. IX, 258 pages. 1978.

Vol. 693: Hilbert Space Operators, Proceedings, 1977. Edited by J. M. Bachar Jr. and D. W. Hadwin. VIII, 184 pages. 1978.

Vol. 694: Séminaire Pierre Lelong – Henri Skoda (Analyse) Année 1976/77. VII, 334 pages. 1978.

Vol. 695: Measure Theory Applications to Stochastic Analysis, Proceedings, 1977. Edited by G. Kallianpur and D. Kölzow. XII, 261 pages. 1978.

Vol. 696: P. J. Feinsilver, Special Functions, Probability Semigroups, and Hamiltonian Flows. VI, 112 pages. 1978.

Vol. 697: Topics in Algebra, Proceedings, 1978. Edited by M. F. Newman. XI, 229 pages. 1978.

Vol. 698: E. Grosswald, Bessel Polynomials. XIV, 182 pages. 1978.

Vol. 699: R. E. Greene and H.-H. Wu, Function Theory on Manifolds Which Possess a Pole. III, 215 pages. 1979.

Vol. 700: Module Theory, Proceedings, 1977. Edited by C. Faith and S. Wiegand. X, 239 pages. 1979.

Vol. 701: Functional Analysis Methods in Numerical Analysis, Proceedings, 1977. Edited by M. Zuhair Nashed. VII, 333 pages. 1979.

Vol. 702: Yuri N. Bibikov, Local Theory of Nonlinear Analytic Ordinary Differential Equations. IX, 147 pages. 1979.

Vol. 703: Equadiff IV, Proceedings, 1977. Edited by J. Fábera. XIX, 441 pages. 1979.

Vol. 704: Computing Methods in Applied Sciences and Engineering, 1977, I. Proceedings, 1977. Edited by R. Glowinski and J. L. Lions. VI, 391 pages. 1979.

Vol. 705: O. Forster und K. Knorr, Konstruktion verseller Familien kompakter komplexer Räume. VII, 141 Seiten. 1979.

Vol. 706: Probability Measures on Groups, Proceedings, 1978. Edited by H. Heyer. XIII, 348 pages. 1979.

Vol. 707: R. Zielke, Discontinuous Čebyšev Systems. VI, 111 pages. 1979.

Vol. 708: J. P. Jouanolou, Equations de Pfaff algébriques. V, 255 pages. 1979.

Vol. 709: Probability in Banach Spaces II. Proceedings, 1978. Edited by A. Beck. V, 205 pages. 1979.

Vol. 710: Séminaire Bourbaki vol. 1977/78, Exposés 507–524. IV, 328 pages. 1979.

Vol. 711: Asymptotic Analysis. Edited by F. Verhulst. V, 240 pages. 1979.

Vol. 712: Equations Différentielles et Systèmes de Pfaff dans le Champ Complexe. Edité par R. Gérard et J.-P. Ramis. V, 364 pages. 1979.

Vol. 713: Séminaire de Théorie du Potentiel, Paris No. 4. Edité par F. Hirsch et G. Mokobodzki. VII, 281 pages. 1979.

Vol. 714: J. Jacod, Calcul Stochastique et Problèmes de Martingales. X, 539 pages. 1979.

Vol. 715: Inder Bir S. Passi, Group Rings and Their Augmentation Ideals. VI, 137 pages. 1979.

Vol. 716: M. A. Scheunert, The Theory of Lie Superalgebras. X, 271 pages. 1979.

Vol. 717: Grosser, Bidualräume und Vervollständigungen von Banachmoduln. III, 209 pages. 1979.

Vol. 718: J. Ferrante and C. W. Rackoff, The Computational Complexity of Logical Theories. X, 243 pages. 1979.

Vol. 719: Categorial Topology, Proceedings, 1978. Edited by H. Herrlich and G. Preuß. XII, 420 pages. 1979.

Vol. 720: E. Dubinsky, The Structure of Nuclear Fréchet Spaces. V, 187 pages. 1979.

Vol. 721: Séminaire de Probabilités XIII. Proceedings, Strasbourg, 1977/78. Edité par C. Dellacherie, P. A. Meyer et M. Weil. VII, 647 pages. 1979.

Vol. 722: Topology of Low-Dimensional Manifolds. Proceedings, 1977. Edited by R. Fenn. VI, 154 pages. 1979.

Vol. 723: W. Brandal, Commutative Rings whose Finitely Generated Modules Decompose. II, 116 pages. 1979.

Vol. 724: D. Griffeath, Additive and Cancellative Interacting Particle Systems. V, 108 pages. 1979.

Vol. 725: Algèbres d'Opérateurs. Proceedings, 1978. Edité par P. de la Harpe. VII, 309 pages. 1979.

Vol. 726: Y.-C. Wong, Schwartz Spaces, Nuclear Spaces and Tensor Products. VI, 418 pages. 1979.

Vol. 727: Y. Saito, Spectral Representations for Schrödinger Operators With Long-Range Potentials. V, 149 pages. 1979.

Vol. 728: Non-Commutative Harmonic Analysis. Proceedings, 1978. Edited by J. Carmona and M. Vergne. V, 244 pages. 1979.

Vol. 729: Ergodic Theory. Proceedings, 1978. Edited by M. Denker and K. Jacobs. XII, 209 pages. 1979.

Vol. 730: Functional Differential Equations and Approximation of Fixed Points. Proceedings, 1978. Edited by H.-O. Peitgen and H.-O. Walther. XV, 503 pages. 1979.

Vol. 731: Y. Nakagami and M. Takesaki, Duality for Crossed Products of von Neumann Algebras. IX, 139 pages. 1979.

Vol. 732: Algebraic Geometry. Proceedings, 1978. Edited by K. Lønsted. IV, 658 pages. 1979.

Vol. 733: F. Bloom, Modern Differential Geometric Techniques in the Theory of Continuous Distributions of Dislocations. XII, 206 pages. 1979.

Vol. 734: Ring Theory, Waterloo, 1978. Proceedings, 1978. Edited by D. Handelman and J. Lawrence. XI, 352 pages. 1979.

Vol. 735: B. Aupetit, Propriétés Spectrales des Algèbres de Banach. XII, 192 pages. 1979.

Vol. 736: E. Behrends, M-Structure and the Banach-Stone Theorem. X, 217 pages. 1979.

Vol. 737: Volterra Equations. Proceedings 1978. Edited by S.-O. Londen and O. J. Staffans. VIII, 314 pages. 1979.

Vol. 738: P. E. Conner, Differentiable Periodic Maps. 2nd edition, IV, 181 pages. 1979.

Vol. 739: Analyse Harmonique sur les Groupes de Lie II. Proceedings, 1976–78. Edited by P. Eymard et al. VI, 646 pages. 1979.

Vol. 740: Séminaire d'Algèbre Paul Dubreil. Proceedings, 1977–78. Edited by M.-P. Malliavin. V, 456 pages. 1979.

Vol. 741: Algebraic Topology, Waterloo 1978. Proceedings. Edited by P. Hoffman and V. Snaith. XI, 655 pages. 1979.

Vol. 742: K. Clancey, Seminormal Operators. VII, 125 pages. 1979.

Vol. 743: Romanian-Finnish Seminar on Complex Analysis. Proceedings, 1976. Edited by C. Andreian Cazacu et al. XVI, 713 pages. 1979.

Vol. 744: I. Reiner and K. W. Roggenkamp, Integral Representations. VIII, 275 pages. 1979.

Vol. 745: D. K. Haley, Equational Compactness in Rings. III, 167 pages. 1979.

Vol. 746: P. Hoffman, τ-Rings and Wreath Product Representations. V, 148 pages. 1979.

Vol. 747: Complex Analysis, Joensuu 1978. Proceedings, 1978. Edited by I. Laine, O. Lehto and T. Sorvali. XV, 450 pages. 1979.

Vol. 748: Combinatorial Mathematics VI. Proceedings, 1978. Edited by A. F. Horadam and W. D. Wallis. IX, 206 pages. 1979.

Vol. 749: V. Girault and P.-A. Raviart, Finite Element Approximation of the Navier-Stokes Equations. VII, 200 pages. 1979.

Vol. 750: J. C. Jantzen, Moduln mit einem höchsten Gewicht. III, 195 Seiten. 1979.

Vol. 751: Number Theory, Carbondale 1979. Proceedings. Edited by M. B. Nathanson. V, 342 pages. 1979.

Vol. 752: M. Barr, *-Autonomous Categories. VI, 140 pages. 1979.

Vol. 753: Applications of Sheaves. Proceedings, 1977. Edited by M. Fourman, C. Mulvey and D. Scott. XIV, 779 pages. 1979.

Vol. 754: O. A. Laudal, Formal Moduli of Algebraic Structures. III, 161 pages. 1979.

Vol. 755: Global Analysis. Proceedings, 1978. Edited by M. Grmela and J. E. Marsden. VII, 377 pages. 1979.

Vol. 756: H. O. Cordes, Elliptic Pseudo-Differential Operators – An Abstract Theory. IX, 331 pages. 1979.

Vol. 757: Smoothing Techniques for Curve Estimation. Proceedings, 1979. Edited by Th. Gasser and M. Rosenblatt. V, 245 pages. 1979.

Vol. 758: C. Năstăsescu and F. Van Oystaeyen; Graded and Filtered Rings and Modules. X, 148 pages. 1979.

Vol. 759: R. L. Epstein, Degrees of Unsolvability: Structure and Theory. XIV, 216 pages. 1979.

Vol. 760: H.-O. Georgii, Canonical Gibbs Measures. VIII, 190 pages. 1979.

Vol. 761: K. Johannson, Homotopy Equivalences of 3-Manifolds with Boundaries. 2, 303 pages. 1979.

Vol. 762: D. H. Sattinger, Group Theoretic Methods in Bifurcation Theory. V, 241 pages. 1979.